Topical Conference: Next-Gen Manufacturing

Held at the 2023 AIChE Annual Meeting

Orlando, Florida, USA
5-10 November 2023

ISBN: 978-1-7138-9293-9

Topical Conference: Next-Gen Manufacturing

Held at the 2023 AIChE Annual Meeting

Orlando, Florida, USA
5-10 November 2023

AIChE
The Global Home of Chemical Engineers

Printed from e-media with permission by:

Curran Associates, Inc.
57 Morehouse Lane
Red Hook, NY 12571

Some format issues inherent in the e-media version may also appear in this print version.

Printed with permission by Curran Associates, Inc. (2024)

AIChE
120 Wall Street, FL 23
New York, NY 10005-4020

Phone: (800) 242-4363
Fax: (203) 775-5177

www.aiche.org

Additional copies of this publication are available from:

Curran Associates, Inc.
57 Morehouse Lane
Red Hook, NY 12571 USA
Phone: 845-758-0400
Fax: 845-758-2633
Email: curran@proceedings.com
Web: www.proceedings.com

TABLE OF CONTENTS

3D PRINTING FUNDAMENTALS AND APPLICATIONS

3D PRINTING NOVEL METHODS AND APPLICATIONS

3D PRINTING OF COMPOSITES

3D PRINTING OF FUNCTIONAL MATERIALS

POLYMERS IN ADDITIVE MANUFACTURING I

CYBERSECURITY AND HIGH-PERFORMANCE COMPUTING IN NEXT-GEN MANUFACTURING

FUTURE OF MANUFACTURING AND EMERGING TECHNOLOGIES

ADVANCES IN SMART MONITORING, OPTIMIZATION AND CONTROL OF PROCESS MANUFACTURING

INDUSTRY 4.0, DIGITAL TWINS, AND DIGITAL TRANSFORMATION

MODELING, OPTIMIZATION, AND CONTROL IN NEXT-GEN MANUFACTURING I

NEXT-GEN MANUFACTURING IN CHEMICAL AND ENERGY SYSTEMS

NEXT-GEN MANUFACTURING IN PHARMA, FOOD, AND BIOPROCESSING I

POLYMERS IN ADDITIVE MANUFACTURING II

4a Additive Manufacturing of Functional Medical Devices for Use in the Clinic

Matthew Becker

The emergence of additive manufacturing has afforded the ability to fabricate intricate, high resolution, and patient-specific polymeric implants. However, the availability of resins with tunable resorption profiles remains a significant hurdle to clinical translation. In this presentation, I will outline our strategies for synthesizing highly functional oligomeric resins that can be photochemically printed into a variety of structures possessing unique mechanical, chemical and degradation properties. I will also describe their use in a number of pre-clinical applications.

4b Single and Multiphase Flows in Well-Defined Periodic 3D-Printed Lattices

Ian Woodward, Yinkui Yu, Catherine Fromen

Additive manufacturing affords precise control over geometries with high degrees of complexity, such as lattices, a class of additive-only metamaterials which have great potential in directing transport phenomena. Realizing applications that leverage aerosol transport through novel lattice materials requires new analytical approaches to predict particle-laden flows through these intricate structures. Compared to a fibrous filter, lattices can achieve length scales that are orders of magnitude larger than most porous media, presenting a challenge to translate existing theoretical frameworks to fully capture the relevant transport phenomena. Conversely, the well-controlled structure of a lattice can be considered a distinct advantage, allowing us to address aerosol physics at representative groupings of unit cells that will enable scaling to the entire part. Advancing fundamental aerosol transport behavior in these newly invented metamaterials has broad implications for a wide range of engineering applications, including separations, filtration, functional coatings, heat transfer, and process intensification. However, our group is specifically motivated to evaluate aerosol-lattice dynamics so that we can integrate them within a working model of the human airway using meaningful physical approximations of airway filtration using the tunable spatial collection afforded by lattice materials. Using both experimental and computational multiphase tools, our group studies single and multiphase transport phenomena within 3D-printed lattices of well-defined structures, ranging from uniform, repeating unit cells, to patterned and directional lattice structures. Measurements of bulk transport through well-defined lattice monoliths is then investigated with computational flow profiles generated in dilute-phase simulations. From these studies, we are able to leverage distinct lattice structures to tune aerosol filtration and spatial deposition, even with lattice monoliths of comparable overall porosity. Using these basic design rules, lattices of different structures were integrated into a model of the human lung, where correctly identified lattices were able to provide deposition profiles aligning with clinical benchmarks. Overall, this work represents an important step forward in building predictive models of aerosol dynamics within well-defined lattice structures for a range of transport applications.

4c Printing 3D Structures with Controlled Chiral Self-Assembly

Mohsen Esmaeili, Kyle George, Nader Taheri-Qazvini, Monirosadat Sadati

The emergence of three-dimensional (3D) printing has advanced the fabrication of on-demand architectures for various applications, from functional devices to organs. However, 3D printing is limited to feature size and lacks the nanoscale design of the constituents vital for functional devices. Herein, we combine the 'bottom-up' approach with a 'top-down' fabrication strategy to program the nanoscale chiral self-assembly of cellulose-based materials in the printed objects. The chiral or helical arrangement, in which layers of nanoparticles/fibers are slightly twisted by a fixed angle relative to the neighboring layers, is known as the origin of the superior fracture resistance of the mineralized chitin in the 'smasher-type' mantis shrimp's dactyl club and the vivid metallic colors in beetles. We design our inks based on cellulose nanocrystals and ether variants of cellulose to built-in nanoscale chiral self-assembly with controlled helical pitch length. The flow-induced alignment intrinsic to the printing process is exploited to direct hierarchical chiral self-assembly in the printed constructs. By tuning the ink formulation, its rheological properties, and printing parameters, we have been able to print architectures with uniformly organized built-in chiral nanostructure. Our biomimetic approach will open the path to developing materials with new optical (dynamic color and photonic properties) and mechanical (toughness, strength) properties, naturally emerging from their nanostructure and transferred into the larger scale printed architectures, expanding 3D printing material technologies well beyond what has been conceived and attempted so far, into a new generation of composite and metamaterials and process design.

4d Developing Blood-Mimicking Fluid Formulations to Match Refractive Index of Compliant SLA 3D-Printed Vascular Model

Elnaz Hosseinzadeh, Beatriz Alejandra Bosques Palomo, José Antonio Lorenzo Oliver, Alan Aguirre Soto

Hemodynamic experiments are crucial for the understanding of how chronic cardiovascular diseases develop, such as aneurysms—a silent often deadly condition [1]. Particle image velocimetry (PIV) is the most common technique used in *in vitro* studies to obtain hemodynamic. However, this technique is heavily limited by the materials used for the creation of anatomical "phantoms" that needs to be optically transparent and for the formulation of blood-mimicking fluids (BMF's). For laser-PIV the main challenge is to match the refractive index of the phantom with that of the BMF. The most common approach for producing flow phantoms is the negative casting in transparent silicone elastomers [2]. It has been determined already that the refractive index of silicone-based materials (~1.41) can be matched with water-glycerol BMF formulations that can be considered as synthetic blood with similar properties to human blood like viscosity and density [3]. Nonetheless, the casting with silicone is a laborious time-consuming process that is not capable of reproducing fine patient-specific anatomical features [4]. Therefore, simplified average-anatomy models and approximate geometries have been mostly used. In contrast, 3D printing enables the fabrication of patient-specific models in a relatively broad range of materials [4]. However, the inherently inhomogeneous surface finish of 3D-printed parts makes it difficult to match the refractive index of the phantom (~1.47–1.51 at 532 nm [5]) with common BMF's [6]. In this study, we investigated the BMF formulations that can match the refractive index of commercial SLA 3d-printable materials [4]–[6]. We selected resins yielding elastomeric 3D printed models that are as close as possible to the silicone-based standards. There are few studies in the literature to measure the refractive index of 3D printing materials and they are focused on hard resins which cannot mimic the realistic behavior of vascular tissue. The interaction of the flexible 3D-printed models with water/glycerol as a base solution, third-component solutions like water/glycerol/sodium iodide and water/glycerol/Urea and water/glycerol/Ammonium thiocyanate was studied. We also designed four-component solutions to achieve higher refractive index by more stable solutions. We develop the first comprehensive calibration to the effect of each component on the refractive index. The 3D printed rectangular strips were immersed in solutions and the transmission were measured via U-Vis spectroscopy quantitatively and with photography to show the potential of the solution qualitatively. The transparency improved with three and four-component solution by 90%. The optical and mechanical properties of each one of the 3D-printed materials was assessed. Viscosity, density, and refractive index were measured for each solution. We report the discovery of what seems to be the best BMFs based on matching the properties to human blood and being capable to match the refractive index of soft SLA 3D printing materials to enable their used in laser-PIV studies for hemodynamic studies. A flow phantom based on this study have been made to study the blood hemodynamics in popliteal aneurysms. The patient-specific 3D-printed phantom was printed in flexible SLA 3D printing materials which has enough optical and mechanical properties for PIV study.

References:

[1] K. M. Saqr *et al.*, "What does computational fluid dynamics tell us about intracranial aneurysms? A meta-analysis and critical review," *J. Cereb. Blood Flow Metab.*, vol. 40, no. 5, pp. 1021–1039, 2020, doi: 10.1177/0271678X19854640.

[2] S. G. Yazdi, P. H. Geoghegan, P. D. Docherty, M. Jermy, and A. Khanafer, "A Review of Arterial

Phantom Fabrication Methods for Flow Measurement Using PIV Techniques," *Ann. Biomed. Eng.*, vol. 46, no. 11, pp. 1697–1721, 2018, doi: 10.1007/s10439-018-2085-8.

[3] M. Y. Yousif, D. W. Holdsworth, and T. L. Poepping, "A blood-mimicking fluid for particle image velocimetry with silicone vascular models," *Exp. Fluids*, vol. 50, no. 3, pp. 769–774, 2011, doi: 10.1007/s00348-010-0958-1.

[4] K. I. Aycock, P. Hariharan, and B. A. Craven, "Particle image velocimetry measurements in an anatomical vascular model fabricated using inkjet 3D printing," *Exp. Fluids*, vol. 58, no. 11, pp. 1–8, 2017, doi: 10.1007/s00348-017-2403-1.

[5] M. S. Song, H. Y. Choi, J. H. Seong, and E. S. Kim, "Matching-index-of-refraction of transparent 3D printing models for flow visualization," *Nucl. Eng. Des.*, vol. 284, pp. 185–191, 2015, doi: 10.1016/j.nucengdes.2014.12.019.

[6] W. H. Ho, I. J. Tshimanga, M. N. Ngoepe, M. C. Jermy, and P. H. Geoghegan, "Evaluation of a Desktop 3D Printed Rigid Refractive-Indexed-Matched Flow Phantom for PIV Measurements on Cerebral Aneurysms," *Cardiovasc. Eng. Technol.*, vol. 11, no. 1, pp. 14–23, 2020, doi: 10.1007/s13239-019-00444-z.

4e Multi-Order Combinatorial Lattices for Spatial Control of Transport Phenomena

Ian Woodward, Catherine Fromen

Additive manufacturing advances have spawned numerous applications across chemical engineering domains, spanning separations, engineered living systems, biomedical devices, and process intensification. These applications frequently leverage additive-only lattice structures as novel substrates to control the relevant transport phenomena through tunable geometric features and functional material properties. However, the potential to use these systems for precise spatial control of phenomena is hindered by choices often made to streamline the design process. In particular, lattice designs tend to use homogeneous parameters and a cubic grid system in which a unit cell is simply stacked along the primary Cartesian axes. Here we present a framework for multi-order combinatorial lattices that facilitates local tunability and inherently complements multi-material additive manufacturing. We report the range of fluid dynamic behavior in lattice columns composed of one-, two-, and three-dimensional alternating combinations of various unit cells with cell-based length scales ranging from 1-5 millimeters. Furthermore, for unit cells like simple cubic and Kelvin, which can be described with characteristic features either at the core or perimeter of the repeat unit, we examine column performance and as-printed dimensional fidelity as a function of standard or alternative unit cell design. In comparing the types of combinatorial structures produced via vat photopolymerization on a Carbon M1, we find results that are consistent with previous reports, and depending on unit cell selection, 2D combinations can achieve lower pressure gradients than 1D or 3D combinations. Furthermore, we find that alternative unit cell designs show nearly identical pressure gradients as standard homogeneous columns, but in combinatorial form with complementary unit cells, there is a statistically significant performance difference between alternative and standard combinations. Finally, we discuss the implications of these findings for structures deviating from cubic grid systems, and we highlight the compatibility of this framework with our previous developments in lattice generation and designs exhibiting directionally dependent flow behavior. Together, these advances capitalize on the design flexibility of additive-only structures to create spatially distributed, dynamic, and responsive phenomena, enhancing the potential of current technologies and enabling new approaches and applications throughout the discipline.

4f A Novel Means of Measuring Key Thermphyscal Properties Needed for Additive Manufacturing in Space

Jason Livesay, Ranga Narayanan, Robert Singiser, Zachary Karpinski, Linnea Helenius, Joshua Fosen

Additive manufacturing (AM) is very important for long-term habitation in space. Processes include directed energy deposition and 3-D printing of metals. Space is a harsh environment where energy use must be vitally minimized and it thus important to use sophisticated process algorithms to remove excess energy use in AM processes. These algorithms must rely on accurate values of thermophysical properties (i.e., surface tension and viscosity) to optimize process conditions for a wide array of liquid metals and alloys. NASA and the National Academies in recent reports have acknowledged knowledge of precise thermophysical properties as essential for materials processing. Measurement of properties at high temperatures is currently accomplished through a small set of methods. A current method employs drop levitation using an electrostatic field and correlating decay rates to properties. The drawbacks are that the drop must be small and levitated by a high DC voltage, inhibiting use in inert gases due to electric arcing. The method outlined in this talk seeks to eliminate drawbacks using a novel means called electrostatic resonance of layers. This will be accomplished using forcing with AC fields across a layer of fluids. Interfacial patterns will be generated as a signal of resonance-induced instability. The critical voltage and the associated patterns at the resonant state can be measured to high accuracy and correlated to surface tension and viscosity via theoretical models.

Support is acknowledged from NSF NNX17AL27G and NASA 80NSSC 21K0352 and NNX17AL27G

4g Fundamental Study on Processing-Structure-Properties Relationships of Acrylonitrile-Butadiene-Styrene (ABS) for Material Extrusion Additive Manufacturing (MatEx AM)

Juhyeong Lee, Jay Park

Today, Acrylonitrile-Butadiene-Styrene (ABS) is known as the most common material for material extrusion additive manufacturing (MatEx AM) due to its well-balanced properties and easy availability. Although a wide range of ABS sources have already been available on the market and there have been many studies using them, very little work has been done on understanding relationship between processing (*i.e.,* polymerization and post processing) – structure (*i.e.,* microstructural multiphase) – properties (*i.e.,* toughness) of ABS for MatEx AM. In this work, these relationships are evaluated using as-printed and annealed Izod impact strength with three or more ABS grades with different characteristics, such as mass or emulsion polymerization, uni- or bimodal distribution of Butadiene (BD) rubber particles in Styrene-co-Acrylonitrile (SAN) Matrix. For as-printed, the order of Izod impact strengths is almost the same as that of bulk (*i.e.,* injection molded part) value. Among them, the interesting thing is that even though bulk values are almost same, ABS grade(s) with bimodal distribution of BD rubber particles has a higher impact strength than the unimodal distribution. On the other hand, it revealed completely different results after post-processing (*i.e.,* annealing). ABS grade(s) made by mass polymerization, which is less as-printed Izod impact strength, has a significant annealing effect on improving toughness and its value had even become almost the same as the bulk value. In addition, ABS grade(s) with bimodal distribution of BD rubber particles has a better annealing effect on its toughness than that with unimodal distribution. Fractographic and rheological characterizations elucidate the processing-structure-properties relationships of ABS for MatEx AM.

187a *Bring Industrial Production to the Point of Need*

Sizhu You

3D printing of composites and metal have been increasingly adopted to bring resilience and flexibility to end-use part production across industries. The adoption is being catalyzed by advances in part performance and reliability, together with emerging platforms enabling part design and qualification. This presentation will highlight a few recent developments in the Markforged ecosystem that further empower FFF (fused filament fabrication) technologies for industrial production. The examples include aerospace qualification of continuous fiber reinforced materials, novel co-parts design for isotropy, prescriptive simulations, and pushing performance boundaries with metal matrix composites.

187b Development of a Custom 3D Printed Paddle Wheel in a Flat Panel Photobioreactor for Improved Algal Growth

Michelle Meagher, Jacob Tamburro, Nanette Boyle

One of the main challenges with using flat panel bioreactors for algal growth is uneven mixing and settling of cells in corners, especially when bubbling alone is used to mix the culture. In order to improve mixing in our flat panel reactor, we designed a custom paddlewheel; paddlewheels are frequently used in outdoor algae raceway ponds to encourage turbulent mixing and we are taking advantage of the same principle for internal mixing in the reactor. The paddlewheel is easily integrated into our PSI FMT150 1 L flat panel photobioreactor and printed on a 3D printer using high temperature poly lactic acid (HT-PLA). With the inclusion of a thermal annealing step, the paddlewheel is autoclavable. Addition of the paddlewheel in the reactor minimized cell settling and improves growth rates with minimal additional cost. We will present our design and data to support improved growth of algae. This work also provides an example of how personalized manufacturing through 3D printing is beneficial in a laboratory environment.

187c Development of Experimentally Validated CFD Model to Study the Drop Dynamics during Inkjet Based 3D Printing of Tablets

Tanu Mehta, Hossain Aziz, Shing-Yun Chang, Anson Ma, Bodhisattwa Chaudhuri

Purpose:

Three-dimensional printing in pharmaceuticals offers multiple advantages such as personalization of medicine, fabrication of complex dosage forms and high drug loading in comparison of conventional dosage forms. Binder jet 3D printing involves jetting of binder solution on top of the powder layer in layer-by-layer fashion. The properties of the formulation are highly dependent upon jetting of the ink/binder solution from the printhead nozzle. In this study a finite volume based computational fluid dynamics (CFD) model has been developed to simulate liquid drop formation in a piezoelectric inkjet printhead. Finite-volume based CFD model works by solving continuity and momentum equations to obtain the fluid flow field in the Eulerian coordinate system in conjunction with appropriate physical models in each control volume. In addition to that, drug solutions utilized by our group for 3D printing of pharmaceutical tablets were also evaluated for the drop properties.

Methods:

In this research, Volume of fluid (VOF) and Continuum Surface Force (CSF) model was used to model the drop formation from the printhead nozzle using commercial software ANSYS Fluent. VOF model is used to track interface and CSF model accounts for the surface tension effects. The piezoelectric nozzle simulated here works by creating pressure inside the nozzle when voltage is applied which consequently leads to drop jetting. The system geometry was created in ANSYS Space Claim with a nozzle diameter of 40μm and a stand-off distance of 5mm (Fig. 1). Meshing was performed using ANSYS meshing (Fig 2), and model setup was performed in ANSYS Fluent. The pressure generation inside the nozzle was implemented by corresponding inlet velocity boundary condition. Experimental determined ink solution properties were used as the input for the CFD model.

Results:

Initially the mesh independence study was performed by analyzing different mesh element sizes and different inlet velocity values to find the optimum mesh element size to avoid the effect of different mesh sizes on the simulation results. Various simulations were performed with different inlet velocities and parameters such as drop diameter, drop velocity, and drop shape were compared with the experimental data for model fluid (Fig. 3) and water as the ink solution. It was observed that as the ink surface tension increases the drop volume and drop velocity decreases. Lower surface tension also increases the number of satellite droplets. It can be observed from the Fig. 4 that 5 mg/ml concentration ink solution with the highest surface tension demonstrated drop volume of 24.10 pL and a drop velocity of 6.95 m/s whereas 40 mg/ml drug concentration ink solution with lowest surface tension showed drop volume of 18.80 pL and drop velocity of 6.32 m/s.

Conclusions:

The drop formation is known to be influenced by the fluid properties such as surface tension, viscosity, and density of the ink solution which were studied using this model for multiple ink solutions used during the experiments. This CFD model will help in evaluating the jettability of different ink solutions thereby helping to improve the printing process. The simulation results and experimental data demonstrated a good agreement indicating the applicability of CFD model to study drop properties during inkjet-based 3D printing process.

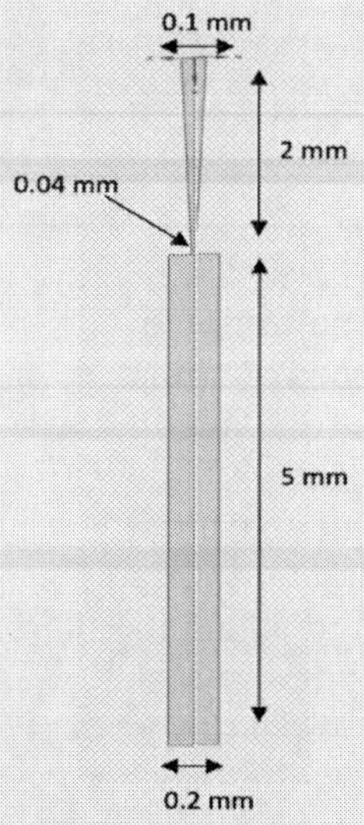

Figure 1: Geometry for piezoelectric printhead

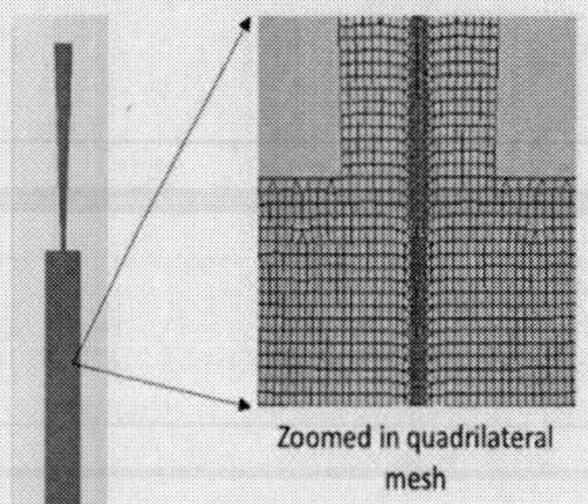

Figure 2: 2D mesh for piezoelectric printhead

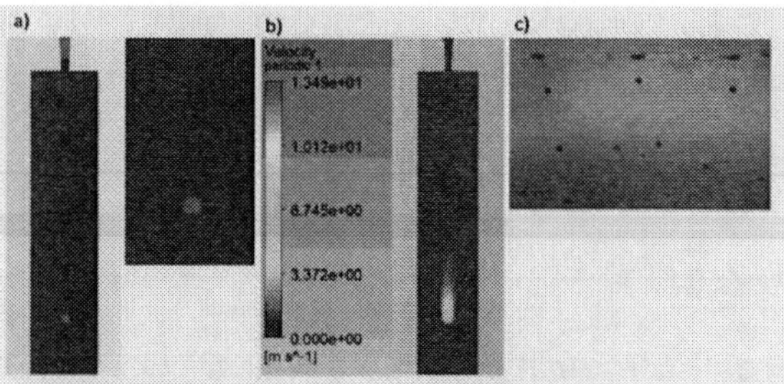

Figure 3: CFD simulation results for degassed water showing:
a) Volume fraction profile
b) Velocity profile
c) Image from drop watcher experiments showing spherical drop formation

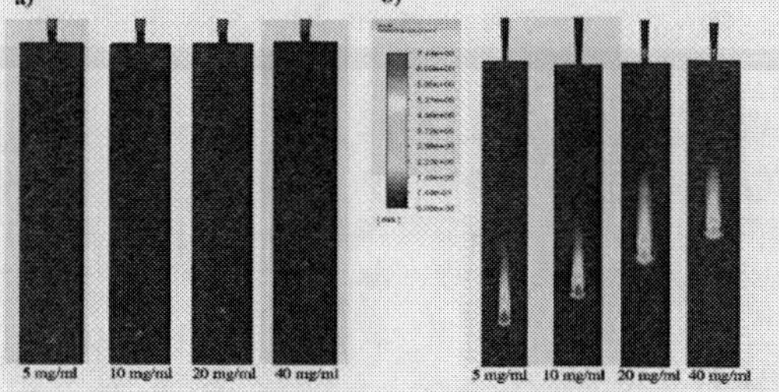

Figure 4: CFD simulation results for ink solutions showing:
a) Volume fraction profile
b) Velocity profile

187d Masked Stereolithography 3D Printing for Rapid Fabrication of Microfluidic Devices with Functional Components and Fluorescence Thermometry

Isteaque Ahmed, Aashish Priye

The rapid prototyping and fabrication of microfluidic devices have been significantly enhanced by the use of 3D printing technologies, such as masked stereolithography (MSLA). This study presents the use of inexpensive desktop SLA printers and the development of a custom UV curable resin to rapidly fabricate microfluidics chips with functional components. Our resin chemistry enables transparent stereolithography prints with tunable mechanical properties and the integration of functional nanostructures in the polymer matrix. We demonstrate the fabrication of unibody microfluidic chips with embedded channels and microscale features, and functional components in a single step print process. Furthermore, we show that by changing the crosslinking density of the resins, we can print flexible and porous components that can serve as microfluidic valves and membranes respectively. Additionally, we explore the feasibility of integrating Rhodamine B dye into the custom resin for non-contact fluorescence thermometry in microfluidic systems, offering a non-invasive and convenient temperature measurement alternative to traditional contact-based methods. We investigated the thermometric properties of Rhodamine B and the compatibility of the dye-resin mixture for photopolymerization and 3D printing. Taken together, this study highlights the benefits of using inexpensive MSLA 3D printing and custom resin formulations for rapid prototyping and fabrication of microfluidic devices with integrated non-contact thermal sensing capabilities, offering significant improvements in efficiency and versatility over conventional photolithography-based methods.

187e Development of a Selective Powder Deposition (SPD) 3D Printer

Guilherme Bertola, Steven Tricarico, Anson Ma

Current powder-based 3D printing methods such as selective laser sintering (SLS) and binder jetting (BJT) create functional gradients by tuning the laser power/ raster speed and the composition and level of saturation of the liquid binders. However, they lack the ability to vary the powder composition *on demand* for creating heterogenous, functionally graded composites. In this presentation, we will share our latest research in developing a novel selective powder deposition (SPD) method capable of selectively depositing various polymer and metal powders without using a liquid carrier or solvent. The SPD method aims to: (i) address unmet needs in mixing powder on demand for SLS and BJT of functionally graded structures, (ii) spare materials by depositing powders only in selected locations that will either form the final part or provide necessary support, thereby minimizing powder usage and reclaim, and (iii) enable solvent-less processing to avoid the need for liquid removal that is not only energy consuming but may also result in unwanted property changes due to a thermal history.

267a Interface-Mediated Mechanics of Highly Filled, Additively Manufactured Polymer Composites

Amy Peterson

Digital light processing (DLP) is a popular method of vat photopolymerization additive manufacturing of thermosetting polymers with high resolution that is well-suited to fabrication of highly filled composites. Interfacial interactions are particularly important in highly filled composite systems given the large surface area of interface per unit volume. In this work, high volume fraction (50 vol.%) glass microsphere-reinforced composites were fabricated via DLP. The effect of surface chemistry and interfacial interactions on the mechanical properties of additively manufactured composites were investigated using two model systems: one where glass microsphere surface chemistry was modified using silane coupling agents, and one where interfacial interactions were mediated with the addition of monomers to the resin system that have previously been shown to improve adhesion through non-covalent interactions. Interfacial interactions between the urethane acrylate-based resin system and glass microspheres were found to dictate the overall performance of the composites. The efficacy of these two approaches are compared with respect to their ability to alter mechanical and thermomechanical properties of the composite.

267b Rheology and Effect of Dispersion on 3D Printing of Chitosan-Graphene-Titanium Dioxide Composites

Michael Bortner, Mariama Alidu, Daniel Alves Heinze

We have demonstrated the ability for Chitosan-Graphene-Titanium Dioxide (CS-G-T) composites for various fields of water remediation to increase removal density of cyanobacteria, specifically Anabaena sp., Synechocystis sp., and Microcystis aeruginosa in harmful algae blooms. Translating these material compositions to 3D printing opens the potential for tailored kinetics and controlled remediation. In this study, we focus on developing CS-G-T polymer composites for direct-ink-write three-dimensional (3D) printing. We enhanced the dispersion of titanium dioxide and graphene by employing probe sonication with appropriate energy adding a cooling medium to prevent water loss through evaporation. We varied the composition of titanium dioxide while maintaining constant chitosan and graphene composition. Dynamic light scattering and scanning electron microscopy were employed to validate dispersion, resulting in a significant improvement in printability due to the enhanced distribution of titanium dioxide and graphene. We further performed torsional rheological measurements to investigate the shear-thinning behavior and yield stress behavior of the inks, critical for enabling extrusion while limiting spread post-deposition. Further, the addition of a processing aid while printing was found to improve printability and decrease the spreading of the polymer composites by forming a hydrogel upon deposition. These results present a promising approach for enhancing the printability of Chitosan- Graphene-Titanium Dioxide (CS-G-T) Polymer Composite for direct-ink-writing 3D printing. This study further contributes to the understanding of the rheological behavior of C-G-TiO_2 composites and their potential applications in water remediation using 3D printed structures.

267c Multichannel Hollow Carbon Fiber Reinforcement in an Epoxy Resin Matrix for Direct Ink Writing of High-Performance Composites

Roneisha Haney

Carbon-fiber-reinforced polymers have been widely used as additive manufacturing materials for high-performance composites. However, as applications within the aerospace and automotive industries evolve, demand for lighter weight materials with feasible processing methods has increased. Herein, we utilized novel hollow carbon fibers (HCFs) with a honeycomb cross-section as lightweight reinforcements for high-performance composites fabricated via direct ink writing. The printability, mechanical performance, and underlying microstructural features of an HCF-based matrix were investigated. According to rheological tests, the pre-printing characteristics of HCF-based inks are similar to those of traditional densified carbon fibers (DCFs). However, mechanical (flexure and tensile) testing found that traditional DCF composites outperform HCF composites. Possible failure mechanisms and variations in mechanical properties were investigated using scanning electron microscopy. Interestingly, this analysis revealed that epoxy infiltrated the entire channel length in the hollow fibers with easy pullout of long, uniform epoxy strands. Although interlocking between fibers and the matrix was expected to provide mechanical improvements, weak fiber–matrix bonding in the fiber interiors resulted in decreased tensile and flexural strengths. These findings demonstrate that appropriate sizing of the surfaces within the HCFs could provide opportunities for interlocking with various matrix systems, including high-temperature resins. Future work entails investigating and characterizing the influence of HCF channels on the resultant strength and exploring novel methods for improving interfacial bonding through modification of the channel surfaces.

267d Development of Fused Filament Fabrication (FFF) Printed Multi-Material Parts with Improved Interlayer Bonding

Daniyal Shoukat, Javaz Rolle, Carson Meredith, Nese Orbey, Jay Park

Poor interlayer adhesion remains a crucial problem in Fused Filament Fabrication (FFF) resulting in meager mechanical properties in 3D printed parts. In this work, protocols developed in our previous work are extended to study the interlayer adhesion between layers of two or more different resins, and between filled and neat layers of the same resin. Effects of layer thickness, volume composition, filler loading, printing temperature and rate on interlayer adhesion are studied. Printed samples are characterized using traditional ASTM methods and High Throughput Mechanical Analysis (HTMECH). Multi-material bilayer films are also prepared using solvent casting and characterized using the HTMECH to correlate off-printer and on-printer behavior. The HTMECH serves as a rapid testing method for high throughput adhesion screening of multimaterial systems reducing testing time associated with standard ASTM methods.

267e Rapid Processing of Carbon Fiber/Polymer Composites Using Electromagnetic Fields

Anubhav Sarmah, Smita S. Dasari, Ethan Harkin, Micah Green

Carbon fiber reinforced composites (CFRCs) are lightweight, mechanically strong, and resistant to chemical and thermal degradation. These properties make CFRCs a popular choice for various applications in military, aerospace, and automobile industries. The processing life-cycle of CFRCs consist of four major stages: fabrication of partially cured prepregs, curing of CFRCs for final applications, repairing of damaged CFRCs, and recycling carbon fibers from end-of-life composites. Conventionally, all these stages rely on large autoclaves or ovens for heat, rendering all these processes extremely energy inefficient. Our group utilizes electromagnetic fields for rapid processing of CFRCs. Carbon fibers rapidly heat up when exposed to non-contact radio-frequency (RF) fields. Thermal energy is generated by the carbon fibers interacting with the RF fields (1–200 MHz range) and drives the crosslinking reaction in the surrounding epoxy to successfully make prepregs of varying degrees of cure, depending on the curing temperature and time. Next, RF fields were also used to further cure these prepregs and fabricate fully-cured CFRCs. Mechanical testing of these composites show that the interfacial modulus and tensile strength of composites cured at the same temperature via RF heating or oven heating show comparable mechanical properties. Composites in the field require repair and maintenance due to wear and tear over their lifecycle or even due to damage in the line of duty. We have explored the possibility of using Dielectric Barrier Discharge (DBD)-generated plasma to carry out such repair processes. The damaged crack was covered by a prepreg patch, and then exposed to an electric field plasma, which heats up the fibers in the prepreg and cures the patch. The patch essentially acts as a structural additive to the damaged part, and mechanical testing showed that composites repaired using plasma or conventional oven heating show similar percent recovery of strength compared to the damaged composite. Finally, direct current (DC) heating was used to recycle continuous carbon fibers from end-of-life composite scrap. Existing methods for CFRC recycling either require large, energy-inefficient ovens, or a mechanical milling process which yields chopped fibers. The electromagnetic heating method we developed allows for recycling of continuous carbon fibers, which have mechanical properties similar to oven-recycled fibers, and we have also successfully demonstrated that recycled fibers can be used to make composites for further use in the industry. We have successfully addressed all the stages of CFRC processing and have replaced conventional methodologies with electromagnetic heating to yield an energy efficient, industrially scalable alternative. Next steps include development CFRC additive manufacturing technologies.

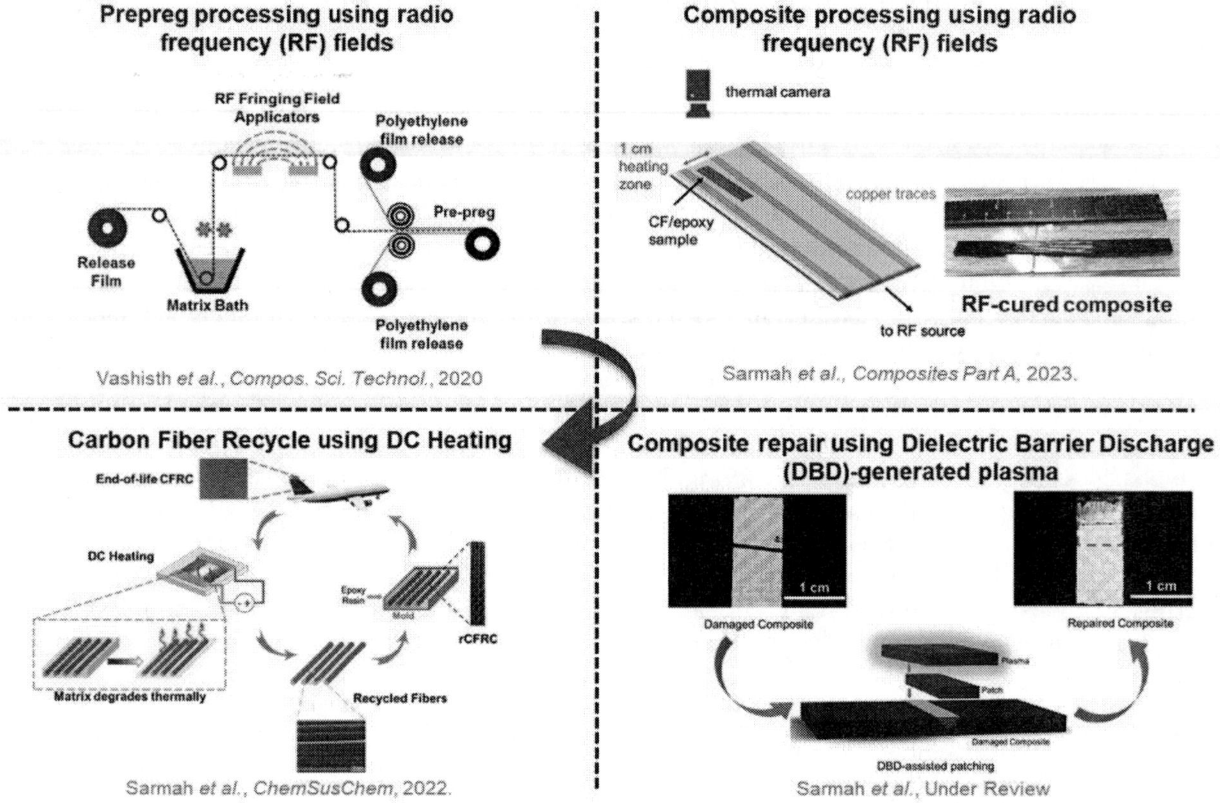
Prepreg processing using radio frequency (RF) fields
RF Fringing Field Applicators
Polyethylene film release
Pre-preg
Release Film
Matrix Bath
Polyethylene film release
Vashisth et al., Compos. Sci. Technol., 2020
Composite processing using radio frequency (RF) fields
thermal camera
1 cm heating zone
CF/epoxy sample
copper traces
RF-cured composite
to RF source
Sarmah et al., Composites Part A, 2023.
Carbon Fiber Recycle using DC Heating
End-of-life CFRC
DC Heating
Epoxy Resin
Mold
rCFRC
Matrix degrades thermally
Recycled Fibers
Sarmah et al., ChemSusChem, 2022.
Composite repair using Dielectric Barrier Discharge (DBD)-generated plasma
1 cm
1 cm
Damaged Composite
Repaired Composite
Plasma
Patch
Damaged Composite
DBD-assisted patching
Sarmah et al., Under Review

267g Enhanced Tunability of Mechanical Properties Using Dual Material Bi-Layered Core-Shell Filaments By Material Extrusion Additive Manufacturing

Nikhil Patil, Kartik Joshi, Kenneth E. Strawhecker, Juhyeong Lee, Timothy Lawton, Eric D. Wetzel, Jay Park

While FFF technique for soft materials is attracting a lot of researchers due to its advantages as a solvent free processing technique, critical processing challenges need to be addressed for fabricating soft material specimens. Present literature fails to provide a robust solution for good print resolution, repeatability of the printed parts, and buckling of filament as critical processing problems. This work focuses on improving structural integrity and mechanical performance of additively manufactured thermoplastic elastomer (TPE) specimens by FFF technique. This approach focuses on improving 3D printing characteristics of TPE (shore hardness 75A) by assembling smaller volume fractions of ABS material, to develop circular core-shell structure of ABS and TPE respectively. Rheological characterizations of feed material using high and low shear viscometry explains melt flow characteristics during extrusion of core-shell filaments as well as wetting characteristic of print layers. Poiseuille flow is used to understand the extrudability of filaments at printer nozzle. Structural integrity and surface uniformity of produced filaments was verified by developing high precision 3D printed complex benchmark geometries. The izod impact and 3-point bending tests analyze the impact strength and flexibility of print structures at various core-shell ratio to achieve tunability of mechanical properties. Prints at 50% ABS and 62% ABS showcase synergy between flexibility and impact properties to develop high modulus and high strength print structures. Enhanced impact strength is due to superior interlayer adhesion between print layers, increasing the resistance to crack propagation. The fracture surface analysis helps to understand the effect of adhesion in print layers to enhance impact characteristics. Decrease in ABS content results in change of izod specimen failure type from ductile failure to elastic failure. In addition to that, increment in fill density showcases enhancement in mechanical properties on mechanical properties of print structure. Microstructures of printed geometries under microscope verified good print uniformity of 3D-printed structures with high packing density within adjacent layers. AFM modulus mapping of print structures explains the formation of localized high and low modulus segments in print structures, enhancing the mechanical properties of resultant print structures.

321a Additive Manufacturing of Carbons from Commodity Plastics

Zhe Qiang, Paul Smith

Additive Manufacturing (AM) is an essential technology revolutionizing how modern manufacturing is approached. Many industries, from medicine to aerospace, are adopting AM to take advantage of its modularity and ease of customization. The application space for AM broadens the need to expand the material design space for AM also grows. While new industrially viable materials systems and AM techniques are rapidly emerging for polymers, metals and composites, this new wave of technology has yet to fully extend to carbon materials. This is despite the exemplary properties of carbon materials and the crucial role they play as additives, fibers, and porous materials in a wide range of applications. Current approaches to perform AM using carbon utilize either highly filled slurries of carbon additives and solvent or polymeric materials that can be pyrolyzed to yield carbon. Despite marked progress in this area, most systems experience challenges with cost, scalability, and dimensional control of the final carbon product. This work seeks to address some of these challenges and establish a scalable and cost-effective way to produce complex, three-dimensional structured carbon materials with good dimensional accuracy. Particularly, it is found that dimensionally accurate structured carbons can be achieved using commercially available materials systems in conjunction with widely available and well understood AM technology.

321b Lightweight Composites: Effect of Shear and Filler Concentration on Alignment, Thermal Conductivity, and Mechanical Properties of Functional Ink

Milan Rede, Marissa Dickerson, Md Alamgir Hossain, Mehul Tank, Gabriela Plautz, Rebekah Sweat, Subramanian Ramakrishnan

Over the past decade, there has been significant interest in the additive manufacturing (AM) of thermosetting resin composites. However, there are issues associated with functionality of the composite, toughness, conductivity, and mechanical properties of the printed structure due to its standard processing techniques. Extrusion-based AM can be used to produce thermoset resin-filler composites with enhanced thermal properties and toughness at loading levels well above the percolation threshold. By optimizing the printing parameters via direct ink writing, lightweight composite materials can be engineered with enhanced properties, so they can be used in applications such as electromagnetic shielding materials, heat sinks, and thermal interface materials because they enhance the thermal coupling between two materials.[1]

In this work, EPON 862 resin was mixed at varied weight concentrations (10%, 11%, 13%, 15%, 16%, 17%, and 18%) of reduced graphene oxide (RGO) fillers to investigate how the macroscopic properties of printed RGO composites are affected at the different weight concentrations, and print speeds (10mm/s and 40mm/s). The rheology of each sample was tested to see the effect of filler concentration on the modulus, viscosity, and yield stress/strain of the samples. The thermal diffusivities and mechanical properties were measured through LFA (Laser Flash Analysis) and DMA (Dynamic Mechanical Analysis) techniques.

The thermal diffusivity of the 13wt% samples printed at 10mm/s were ~0.80 mm2/s at room temperature and 0.66 mm2/s at 100C. When printed at 40mm/s, they were 0.87 mm2/s at room temperature and 0.70 mm2/s at 100C. For the samples at 18wt% printed at 10mm/s, the thermal diffusivities were ~1.09 mm2/s at room temperature and 0.89 mm2/s at 100C and, and when printed at 40mm/s, the thermal diffusivity was 1.22 mm2/s at room temperature and 1.01 mm2/s at 100C. These results indicate as print speed increases, the thermal diffusivity and conductivity increases. This increase can be attributed to the shear-induced alignment of the particles that are extruded on the printing substrate. DMA analyses also showed an increase in Tg (glass transition temperature) at 40mm/s compared to 10mm/s print speed, due to the alignment of the filler particles reinforcing the strength of the material by restricting the polymer chains at higher temperatures. Further experiments are being done to measure the mechanical properties of the samples and investigate the effect of shear on the tensile strength of the composite.

[1]Haney, R. et al. Printability and Performance of 3D Conductive Graphite Structures. Additive Manufacturing 2021.

321c Dynamic, Remote-Controllable Electroactive Hydrogel Waveguide Architectures.

Oscar Alejandro Herrera Cortes Sr.

Electroactive light-guiding structures are generated inside of a hydrogel within seconds using self-trapped visible beams from light-emitting diodes (LEDs) elicited when photo-induced refractive index changes counteract the natural divergence of light. Due to irreversible refractive index changes experienced by photoinduced chemical reactions, these self-trapped beams permanently inscribe waveguides along their paths. The waveguides' orientation, motion and ensuing direction of their light output can be precisely and remotely controlled through external electrical fields. By applying and varying external electric fields, we can then dynamically control the bending, angular orientation and rotation (up to 360°) of these pliant light-guiding structures. This allows precise, remote control of the waveguided light output.

321d Radio Frequency-Assisted Curing of on-Chip Printed CNT/Silicone Heatsinks Produced By Diw 3D Printing

Thang Tran, Anubhav Sarmah, Ethan Harkin, Smita S. Dasari, Kailash Arole, Matthew Cupich, Aniela Wright, Hang Li Seet, Sharon Mui Ling Nai, Micah Green

With the rapid development of high-power integrated electronic devices in recent years, many polymer-based thermal management devices have been developed to address the overheating and improve the performance, reliability, and lifetime of electronic devices. Here, we demonstrate carbon nanotube (CNT)/silicone heatsinks by Direct Ink Writing (DIW) 3D printing and applied Radio Frequency (RF) heating for *in situ* curing. CNTs were used as a thermal conductive nanofiller and a rheological modifier to improve thermal conductivity and printability of the CNT/silicone composite inks, respectively. Additionally, CNTs are also a RF susceptor, so the integration of CNTs into the silicone matrix allowed rapid out-of-oven curing by applying RF heating. Specifically, at CNT loading of 14 wt.%, the CNT/silicone heatsinks could be printed and RF cured with shape retention; the heatsinks' performance was comparable to that of commercial heatsinks. More importantly, the CNT/silicone heatsinks could be printed and cured directly on a chip attached to a circuit board, potentially eliminating the use of thermal interface materials and the step of heatsink installation. The results suggest that the DIW combined with RF assisted curing process can be an effective approach to fabricate high-performance polymer-based heatsinks with reduced production time and energy requirements.

378a Designing Polymer Nanocomposites for Functional Additive Manufacturing

Weinan Xu

Additive Manufacturing or 3D printing of polymers has progressed tremendously in the past decade with various technologies including photopolymerization, extrusion, and powder-based methods. Beyond one component 3D printing, additive manufacturing of polymer composites, however, still faces major challenges such as incompatible processing condition, low printing quality, and poor internal structure control. To address those issues, we have been working on designing polymer composites with unique physical properties and high 3D printability for functional additive manufacturing.

In the first part, I will discuss our recent progress in designing thermoplastic elastomers (TPEs) and their blends for additive manufacturing of soft, stretchable, and biocompatible structures. Polyisobutylene (PIB)-based TPEs including poly(styrene-b-isobutylene-b-styrene) (SIBS) copolymers and PIB-based polyurethanes have been successfully used for fused filament fabrication (FFF) 3D printing. The microphase-separated structure of TPEs can be precisely tuned by adjusting the hard to soft block ratio and blending with rigid homopolymers. Butyl rubber and its nanocomposites have been successfully 3D printed with solvent-assisted direct ink writing (DIW) by adjusting the rheological, mechanical, and electrical properties with nanoclay and graphene. Soft and stretchable strain sensors based on the 3D printed structures were also demonstrated.

In the second part, I will discuss our recent work in designing polymer composite particles for selective laser sintering (SLS) 3D printing. SLS has several unique advantages including the ability to fabricate complex geometries without support structures, highly isotropic properties of the printed parts, batch production of multiple parts in one printing. However, currently there are very limited types of polymer powders available for SLS, with over 95% being polyamide. To address this limitation, we have developed a hybrid approach for the scalable synthesis of polyolefin composite particles, which exhibit narrow size distribution, regular shape, and excellent SLS 3D printability. The processing-structure-property relationship of the polymer composite powders and 3D printed parts was also systematically elucidated.

Key references:

Shen, N.; Duan, L.; Yang, M.; Liu, S.; Xu, W. Solvent-Assisted 3D Printing of Butyl Rubber and Its Hybrid Materials. *MRS Commun.* **2022**, *12*, 597–602.

Shen, N.; Bu, J.; Prevot, M. E.; Hegmann, T.; Kennedy, J. P.; Xu, W. Macromolecular Engineering and Additive Manufacturing of Polyisobutylene-Based Thermoplastic Elastomers. II. The Poly(styrene-b-isobutylene-b-styrene)/Poly(phenylene oxide) System. *Macromol. Rapid Commun.* **2022**, 2200109.

Shen, N.; Liu, S.; Kasbe, P. S.; Khabaz, F.; Kennedy, J. P.; Xu, W. Macromolecular Engineering and Additive Manufacturing of Poly(styrene-b-isobutylene-b-styrene) (SIBS). *ACS Appl. Polym. Mater.* **2021**, *3*, 4554–4562.

Liu, S.; Kasbe, P. S.; Yang, M.; Shen, N.; Duan, L.; Mao, Y.; Xu, W. Intimately Bonded 2D Materials and Responsive Polymer Brushes for Adaptive Nanocomposites. *Polymer* **2020**, *210,* 123033.

378b Dynamic Covalent Photopolymer Networks for Additive Manufacturing

Alexa Kuenstler

Covalent adaptable networks (CANs) leverage exchangeable bonds to enable reconfigurability and recyclability of cross-linked polymer networks. Recently, this paradigm has been extended to additive manufacturing, where triggering dynamic chemistry can be leveraged to both extend the life of printed parts through self-healing or plasticity, while also facilitating life cycle control through exchange-mediated degradation. In this work, we introduce a method to incorporate associative exchange chemistry within photopolymer resins that are amenable to stereolithographic additive manufacturing. We show that through judicious selection of resin components, parts can be printed that range from tough thermoplastic-like materials to soft, resilient elastomers. Significantly, incorporation of dynamic bonds facilitates degradation into low viscosity oligomers under mild conditions. These oligomers can subsequently be re-incorporated into new printed parts of targeted composition, mechanical properties, and function. Finally, we demonstrated that dynamic bonds also engender advanced capabilities within printed parts, such as shape memory through exchange-induced plasticity and self-healing of damaged components.

378c Dual-Wavelength Digital Light Processing Toward Multimaterial 3D Printing

Kyle Chin, Grant Ovsepyan, Andrew Boydston

Additive manufacturing (AM), commonly referred to as 3D printing, is a powerful method for controlling the placement of molecular building blocks—providing increased flexibility in the design, function, and manufacturing of objects. In many AM techniques, object formation is achieved by material deposition directly to specified locations. In contrast, vat photopolymerization (VP) AM builds objects through light-patterned curing of a homogeneous liquid from a material reservoir. This photopatterning approach gives rise to high resolution (~100 μm) prints but makes multimaterial VP challenging. Engineering approaches, such as swapping of resin vats or fluidic devices, have been the predominate solutions. However, they have several drawbacks including increased print times, poor interfacial adhesion, increased manufacturing complexity, and resin contamination. In this talk, I will discuss a chemistry-oriented approach to multimaterial VP that incorporates orthogonal photochemical pathways into resin design. Toward this aim, I will describe the design of various resin formulations that achieve controlled stiffness, disparate solubility, and localized acidity toward creating stimuli responsive materials, dissolvable support structures, and multicolor architectures.

378d Additive Manufacturing of Micro-Architected Metals Via Hydrogel Infusion

Max A. Saccone, Rebecca A. Gallivan, Kai Narita, Daryl W. Yee, Julia R. Greer

Metal additive manufacturing (AM) has emerged as a uniquely powerful tool to produce complex and high-performance parts with applications from the aerospace to biomedical fields. Most existing metal AM techniques use heat to define part shape via melting or sintering of metal powders. In contrast, we report an AM technique, coined hydrogel infusion additive manufacturing (HIAM), that produces metals and alloys with microscale resolution via vat photopolymerization[1]. To fabricate micro-architected metal structures, we infuse 3D printed hydrogels with aqueous metal precursors, then calcine and reduce the infused hydrogel scaffolds to create miniaturized metal replicas. Unlike existing vat photopolymerization strategies, which incorporate target materials or precursors into the photoresin during printing, HIAM does not require re-optimization of resins and curing parameters for different materials, enabling quick iteration, compositional tuning, and the ability to fabricate multimaterials. This talk will elucidate relationships between processing, microstructure, and mechanical properties of HIAM metals, and highlight the wide-ranging versatility of achievable materials and microstructures, which include copper, nickel, silver, cobalt, cupronickel alloys, high entropy alloys, tungsten, and metal multi-materials.

(1) Saccone, M. A.; Gallivan, R. A.; Narita, K.; Yee, D. W.; Greer, J. R. Additive Manufacturing of Micro-Architected Metals via Hydrogel Infusion. *Nature* **2022**, *612*, 685–690. https://doi.org/10.1038/s41586-022-05433-2.

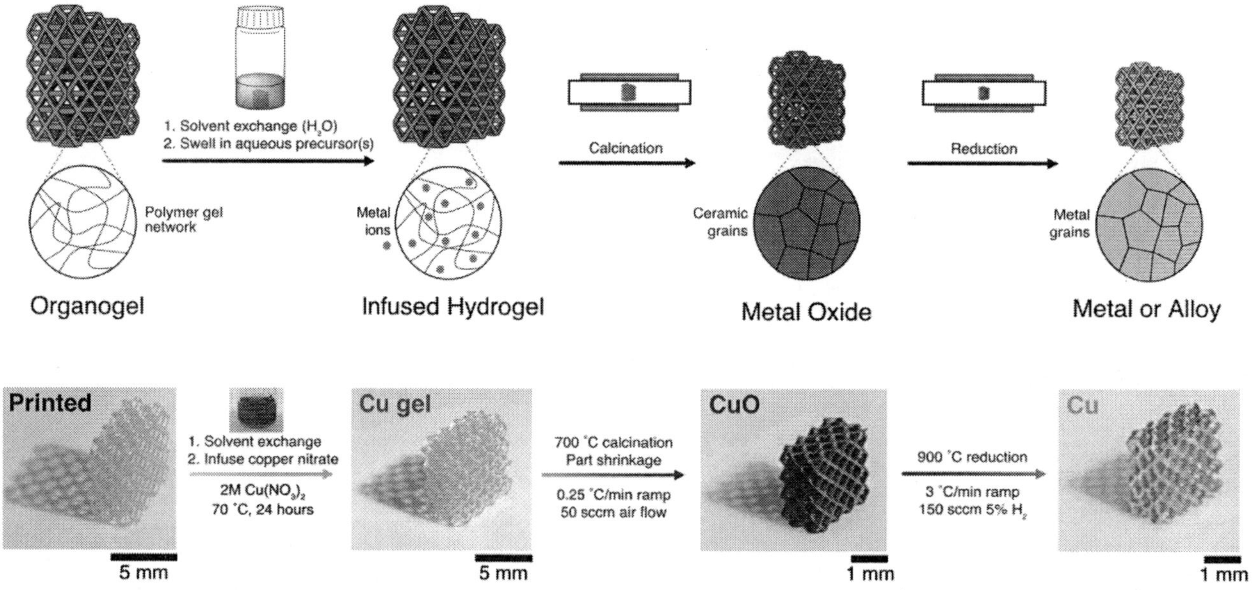

378e 3D, Shape-Specific, Scalable, Micro-Fabricated Particle Production Via Roll-to-Roll Printing and Continuous Liquid Interface Production (r2rCLIP)

Jason Kronenfeld, Lukas Rother, Max Saccone, Maria T. Dulay, Joseph M. DeSimone

Particle fabrication on the hundreds of micrometers to nanometer scale has gained interest in the past several decades with a broad potential range of applications from advanced biomedical devices and drug delivery to intrinsic materials interests in microfluidics, granular systems, and abrasives. Well-studied approaches exist including grinding, emulsification, precipitation, and nucleation-and-growth; however, these techniques inherently yield limited control over geometric complexity and uniformity. To address these shortcomings, several new approaches have arisen including direct lithography, single-step roll-to-roll soft lithographic approaches like PRINT (Particle Replication in Nonwetting Templates), and multi-step lithography approaches like SEAL (StampEd Assembly of polymer Layers) with demonstrated scalability and application in bioanalytics and drug delivery. Further fabrication control is accessed through additive manufacturing (AM), enabling industrially scalable complex geometries and tunable mechanical properties. Particle fabrication via flow lithography and two-photon polymerization (TPP) yield high throughput and high spatial resolution, respectively, but have proven limited in concurrently achieving complex geometries, timely production, and regio-specific chemical composition. High-resolution continuous liquid interface production (CLIP) enables polygeometric, permutable, material versatile particle production at an ultimate feature resolution of 1.5 μm. We report a novel, roll-to-roll high-resolution CLIP (r2rCLIP) process for scalable, shape-specific, micron-scale particle production. The r2rCLIP system is demonstrated through increasing geometric complexity, scalability, and intra-print permutability to form moldable, multistep moldable, and non-moldable geometries at ≤ 200 μm unit size. r2rCLIP is shown to achieve versatility through a 200 μm unit size range and geometric variation within a singular print; geometric complexity up to non-moldable, unreachable hollow inner cavities; and fabrication speeds of up to 527, 200 μm unit particles per minute (9 particles / second). Gram-scale, complex geometry, particle fabrication with inherent amenability to a wide variety of resin chemistries can thus be achieved on the day-scale for translatable biomedical and analytical applications.

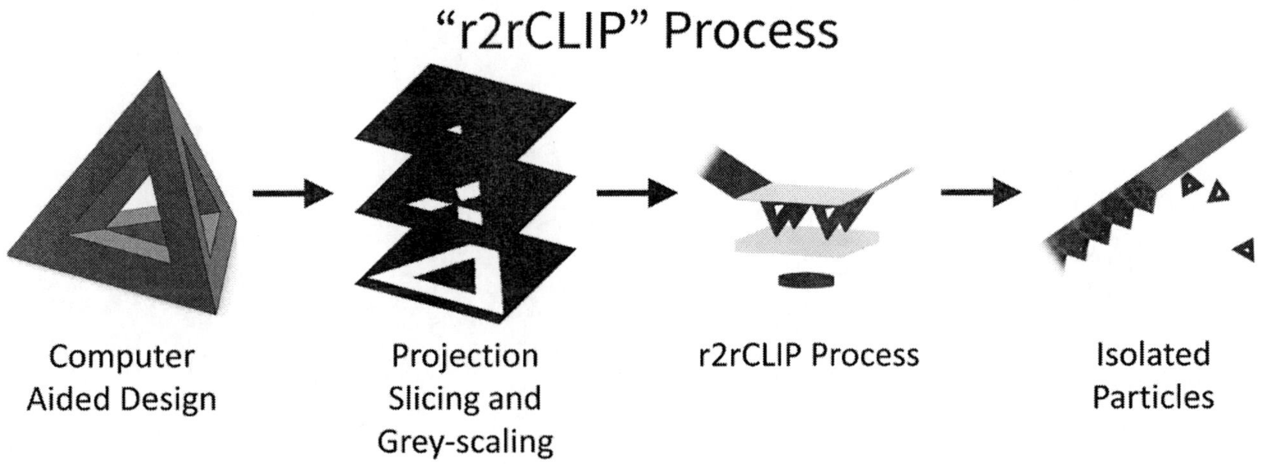

378f Enhancing Functional Properties through a Layered Hierarchy of Nanocomposites through 3D Printing

Dharneedar Ravichandran, Kenan Song

Nanoparticle-reinforced composites are known for their anisotropic behavior and improved physical and chemical properties, but optimal performance requires uniform dispersion, strong particle-matrix interaction, and high particle alignment. While conventional manufacturing excels in uniform dispersion and strong interactions, high alignment of particles is difficult to achieve. 3D printing has the potential to achieve precise nanoparticle alignment of various dimensions and provide customizable structures with reduced lead time and material waste. In this study, a novel nozzle design inspired by natural layered architectures is introduced, which enables the production of composites with a layered structure. This nozzle can simultaneously process two immiscible feedstocks and arrange them in an alternating ABAB fashion, with the number of layers adjustable to influence the alignment and packing coefficients of nanoparticles and polymers. Integrated into the multiphase direct ink writing mechanism, the nozzle allows for fiber fabrication and micropatterning for electronic devices. Various combinations of polymers and nanoparticles can be used for applications like textile fabrics, thermal management devices, smart materials, and soft robotics. The approach combines the advantages of 3D printing, high nanoparticle alignment, and modular design, providing new possibilities for the fabrication of advanced composites with tailored properties.

378g Self-Powered Thermoelectric Biosensor Manufactured By 3D Printing

Yuxiang Zhu, Kenan Song

Applications of thermoelectric generators(TEGs)are limited due to the rigidness of their materials and lack of elaborately designed structures. To harness body temperature as a stable energy source, a gill-mimicking structured TEG is developed, allowing heat to flow through the gills and building up a relatively high temperature difference. Owing toitscustomizability,3D printing plays a significant role in designing and manufacturing the unique-structured generator. The printed structure circumvents the rigid nature of conjugated polymers and fits well on human skin, making the most of body temperatures. The TEG can generate voltages from various complex surfaces, including a human forearm, a hot water bottle, and a car roof. After connecting to a strain switch, biosignals such as joint movement and respiratory rate can be detected without external power sources, which exhibits its potential in long-term health monitoring and smart wearables. The study demonstrates the importance of additive manufacturing in the field of biosensors. Besides the customizability, 3D printing is expected to contribute to higher biocompatibility, miniaturization, tunable porosity, and multifunctionality as well

134a 3D Printing of 2D Materials for Optimized Electrochemical Performance

Marcus Worsley

Electrochemical energy storage (EES) and conversion devices (e.g. batteries, supercapacitors, and reactors) are emerging as primary methods for global efforts to shift energy dependence from limited fossil fuels towards sustainable and renewable resources. These devices, while showing great potential for meeting some key metrics set by conventional technologies, still face significant limitations. For example, an EES device tends to exhibit large energy density (e.g. lithium-ion battery) or power density (e.g. supercapacitor), but not both. This inability of a single device to simultaneously achieve both metrics represents a major obstacle to widespread adoption of EES devices. Similarly, many catalytic processes depend on expensive platinum group metals (PGM) because non-PGM materials have poor performance or durability. Though the integration of 2D materials (e.g. graphene, dichalcogenides, MXene, etc.) into electrochemical devices has yielded some exciting results towards tackling these issues, significant improvements are still needed. One approach to optimizing the performance of these devices is to focus on one of the fundamental processes that occur in these systems: mass (or charge) transport. The efficient transport of ions within EES and conversion devices is critical to realizing better performance and durability. The pore structure of the electrode is a key factor in determining this transport phenomena, but in many cases, engineering the pore structure in a highly deterministic fashion can be challenging. In this work, we explore a number of additive manufacturing methods (e.g. direct ink write, projection microstereolithography, etc.) to engineer the pore structure of device electrodes. We also determine effective electrode geometries using both simple theory and topology optimization techniques. The topology optimization couples the solution of the forward electrochemical problem over the full electrode domain with gradient-based optimization. The output of our code is a threedimensional CAD representation which optimizes over specific performance metrics and which can be used to print functional electrodes. This work provides a systematic path toward automatic design and fabrication of engineered electrodes with precise control over the fluid and species distribution.

This work was performed under the auspices of the U.S. Department of Energy by Lawrence Livermore National Laboratory under Contract DE-AC52-07NA27344.

134b Design, Synthesis, and Catalytic Testing of 3D-Printed Fischer-Tropsch Catalysts

Luis C. Caballero, Joseph Brindle, Sayed Abu Sufyan, Nathan Ramey, Swomitra Mohanty, Michael Nigra

Fischer-Tropsch synthesis (FTS) is a strategically significant reaction to produce high-value liquid chemicals and hydrocarbon fuels from synthesis gas. Optimizing industrially favorable processes, such as FTS, remains challenging in catalysis. As a complex surface polymerization reaction, FTS highly depends on process conditions and catalyst behavior. Its product distribution and deactivation rates have been identified as functions of catalyst temperature, with hot spots leading to undesirable sintering and rapid deactivation. Historically, the preferred support for Fischer-Tropsch catalysts has been titanium dioxide (TiO_2), with metals such as iron, cobalt, and ruthenium comprising the active phase. In this work, we design, synthesize, and test novel FTS catalysts, focusing on designing support materials with improved heat transfer properties. A method using additive manufacturing (3D printing) based on direct metal laser melting (DMLM) is introduced alongside extensive CAD prototype modeling to develop thermally-conductive catalytic supports. A two-step electrochemical anodization process is used to develop TiO_2 nanotube arrays on the Ti-based supports. Different annealing processes for optimizing these nanotubular arrays are also assessed. Bimetallic FeCo-based catalysts (2.0 wt. %) are synthesized on the TiO_2 nanotubular arrays by a modified incipient wetness impregnation method.

The performance of the catalysts are measured using online gas chromatography-mass spectroscopy. Field-emission scanning electron microscopy, X-ray diffraction, and energy-dispersive X-ray spectroscopy assisted in characterizing their morphology, elemental composition, and microstructures at different stages. The results obtained by the characterization of these materials show that the methods presented in this work for synthesizing novel FTS catalytic supports are successful. The bimetallic FeCo catalysts exhibit catalytic activity, and their performance is correlated to their structure. By presenting these novel methods for catalyst synthesis, this work poses a pioneering approach to three-dimensional structured catalyst design to optimize FTS.

134c 3D Printed Flow-Directing Electrodes with Nano-/Micro-Scale Porosity for Controlling Transport and Reactions

Max A. Saccone, Xi Chen, William Tarpeh, Joseph M. DeSimone

Systems that combine optimize transport and reaction parameters are crucial for chemical process intensification. In this talk, we discuss recent progress towards additively manufactured electrode materials that control flow fields in electrochemical systems with forced convection, coined "flow-directing electrodes." We show how 3D printed micro-scale architected structures formed from metal-coated polymers, pure metals/oxides, or pyrolytic carbon can tune flow properties and prevent fouling in electrochemical systems while simultaneously providing pathways for electron transfer. We additionally show how to create multiscale hierarchy through the incorporation of nano-scale porosity in these systems via nanomaterial templating (e.g., ZnO nanorod templating in Fig. 1a) to tune specific surface area and permeability. We demonstrate the utility of such flow-directing electrodes for applications ranging from electrochemical remediation of trace aqueous pollutants (e.g., per- and poly-fluoroalkyl substances (PFAS), nitrates, or phosphates) where a nano-porous layer could act as a size-exclusion membrane (Fig. 1b) to design of electrodes for redox-flow batteries (Fig. 1c) where transport of working ions and redox-active material is maximized but transport and crossover of other species such as electrolytes and mediators are minimized. Flow-directing electrodes represent a paradigm shift in electrochemical reactor design in which high resolution vat photopolymerization printing techniques are used to create co-designed structures with complex geometry using electrically conductive functional materials—all leading to the ability to explore a wide landscape of previously impossible device designs.

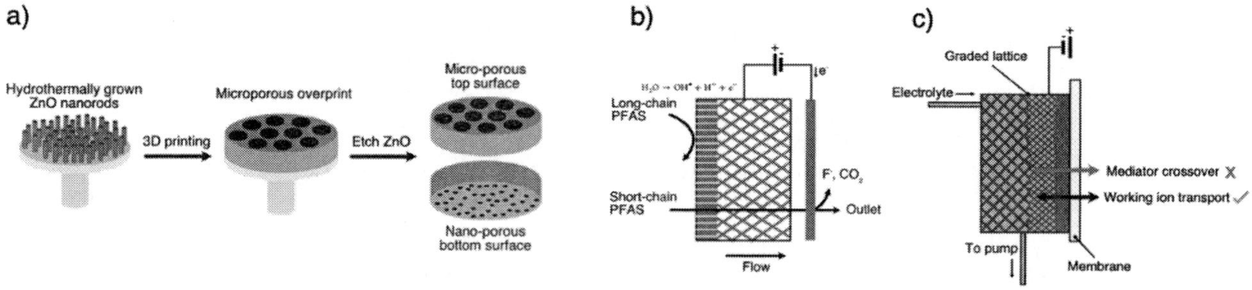

Figure 1. a) ZnO nanorod templating strategy to produce nanoporosity, b) schematic of a flow-directing electrode for electrochemical remediation of trace pollutants such as PFAS, c) schematic of a flow-directing electrode used in a redox-flow battery.

134d Innovations in High Performance Thermal Systems through Additive Manufacturing

Seyed Alireza Rozati, Anju Gupta

It is desirable to enhance the performance of pool boiling systems to enable higher heat dissipation and to reduce the cost of the thermal equipment including, but not limited to distillation reboiler, desalination, steam generators, high-powered electronic systems, and refrigeration. Phase change heat transfer by boiling can be enhanced by introducing characteristics in solid-liquid interface through multidimensional coatings and geometries. For various manifestations, the pool boiling performance of a surface is dictated by higher critical heat fluxes (CHF) and heat transfer coefficients (HTC) at lower wall superheat. In this talk, I will describe the role of additive manufacturing in enhancing the pool boiling performance. Specifically, I will present our recent work on Fused Filament Fabricated (FFF) 3D printed copper with combined geometry and porosity that yielded higher heat transfer coefficients. Additionally, I will discuss our efforts of creating geometries through magnetically aligned metal inks that achieved 49.9% in CHF and 105% in HTC compared to plain copper surface. I will discuss the new fundamental mechanisms correlating wickability and altered three-phase contact angle that attributed to these enhancements.

134e Ionically Conductive 3D Printed Polymer Membranes As a Safe Alternative to Traditional Liquid Electrolyes in Lithium Sulfur Batteries.

Wissam Fawaz, K.Y. Simon Ng

The development of a safe, all solid-state lithium metal (SSLM) battery, that can rival the performance of its flammable, liquid-based electrolyte counterpart is regarded as the next step in the realm of energy storage research. Currently employed liquid solvents in commercial lithium-ion batteries across industries are toxic, polar, flammable, and present a challenge to the ever-expanding milieu of electrification of transportation. A key hurdle to overcome in order to herald SSLM batteries into mass usage is the low ionic conductivity and high interfacial resistances of solid electrolytes.

We herein present a method of surface engineering of a novel crosslinked polymer, for use as solid electrolyte for lithium sulfur batteries, through stereolithography of a mixture of acrylates, polyethene oxide, and lithium salt with appropriate pultruding topography designed to minimize impedance to ionic transfer across battery layers. To aid in conductivity, appropriate amounts of ionic liquid and silicon oxide are added to the electrolyte solution before printing the layers. Our team studied the effect of the electrolyte surface design, ionic liquid and filler content as well as curing time on the electrochemical performance of the formed membranes.

To that effect, cells were assembled with flat electrolyte layers with varying percentages of ionic liquid and silicon oxide content and their ionic conductivity gauged through impedance spectroscopy as well as electrochemical charge / discharge cycling. It was found that both ionic liquid and fillers are integral for the electrolyte membrane performance, as cells lacking either component fail all capacity tests. We asserted that cells fabricated with 3D printed electrolyte membranes in a thin layered setup can deliver a competitive impedance spectroscopy directly proportional to the presence of both ionic liquids and silicon oxide fillers.

134f CFD Evaluation of Six Structured Supports Manufactured By Robocasting As Candidates for High-Temperature Catalytic Applications

Mamoun Al-Rawashdeh, Shaikha AL-Kuwari, Fawziya AL-Darwish, Konstantinos Kakosimos

Heterogenous catalysts are the backbone of the chemical industry. Extrusion is one of the main industrial methods used to shape many of the established commercial catalysts. Commercial catalysts are available in different shapes and sizes that include cylindrical, trilobed, and quadrilobed for pelleted type catalysts, while for structured catalysts, the monolith type is the most established one. Additive Manufacturing (AM) offers a significant potential to explore new unconventional catalyst designs to achieve process innovation. Many additive manufacturing techniques are recently developed for a wide range of different materials and applications. Of these different AM techniques, robocasting is a very promising technique to manufacture heterogeneous catalysts [1]. It is based on the well-established extrusion process but utilizes the principle of additive manufacturing in which extrusion is made in a layer-by-layer structure. In this work, new unconventional 3D-printed catalyst designs that are not possible to produce using conventional extrusion techniques are benchmarked against the well-established monolith catalyst structure. The basic idea for these new designs is to rotate the new layer as the extrusion is progressing. Six (6) new designs are created with rotational angles of 0°, 9°, 18°, 30°, 45°, and 90°. To prove the manufacturing feasibility of these geometries, the designs were fabricated using a benchtop robocasting machine. The performance difference between these catalyst geometries was numerically investigated using Computational Fluid Dynamics (CFD) using COMSOL. This is a good starting approach to evaluating novel catalyst designs [2]. The momentum equation was used to simulate the velocity profile and the pressure drop between these geometries. Steam methane reforming to produce synthesis gas is used as an application case to demonstrate the difference between these geometries. A comparative table is created to benchmark these geometries. Geometry D-0° gave the lowest pressure drop, while D-18° gave the highest. Geometry D-18°, D-30°, and D-45° gave the highest degree of radial mixing. Geometries D-90° and D-9° gave a compromise between lower pressure drop with some degree of radial mixing. This work is a step in the direction of exploring how AM can generate tailored and novel catalyst-shaped designs.

(1) Lawson, S.; Li, X.; Thakkar, H.; Rownaghi, A. A.; Rezaei, F. Recent Advances in 3D Printing of Structured Materials for Adsorption and Catalysis Applications. *Chem. Rev.* **2021**, *121* (10), 6246–6291.
https://doi.org/10.1021/ACS.CHEMREV.1C00060/ASSET/IMAGES/LARGE/CR1C00060_0041.JPEG.

(2) Asif, M.; Grande, C. A. TPMS Contactors Designed with Imprinted Porosity: Numerical Evaluation of Momentum and Energy Transport. *Ind. Eng. Chem. Res.* **2022**, *61* (50), 18556–18566.
https://doi.org/10.1021/acs.iecr.2c03384.

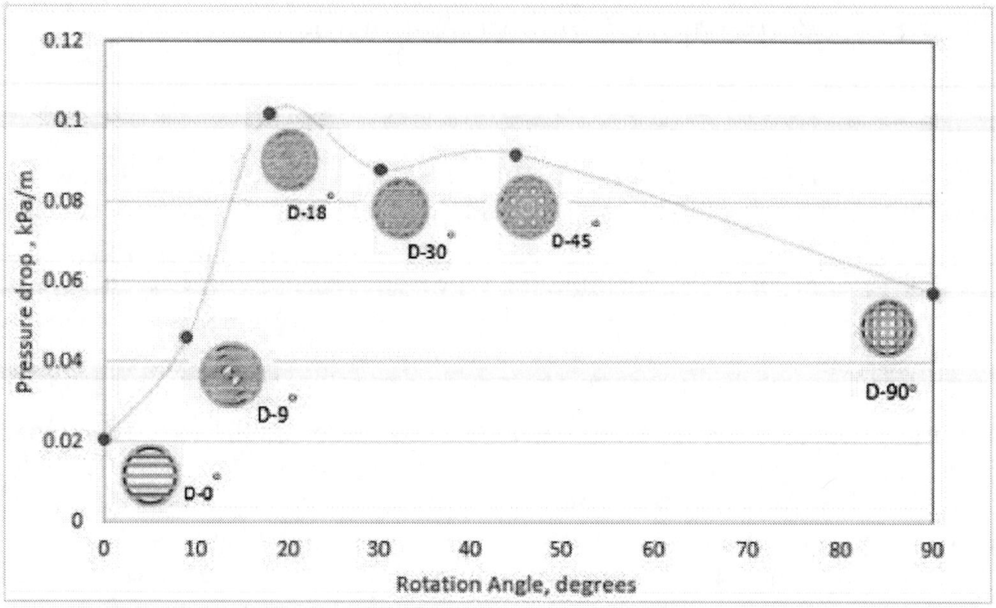

Figure 1. Effect of varying angle on the pressure drop for different rotating angle geometries at a fixed Re = 300

135a Real-Time Feedback Control of an Atomic Layer Etching Spatial Reactor

Feiyang Ou, Matthew Tom, Henrik Wang, Gerassimos Orkoulas, Panagiotis Christofides

Real-Time Feedback Control of an Atomic Layer Etching Spatial Reactor

Matthew Tom[1], Henrik Wang[1], Feiyang Ou[1], Gerassimos Orkoulas[3], and Panagiotis D. Christofides[1,2]

1. Department of Chemical and Biomolecular Engineering, University of California, Los Angeles, CA
2. Department of Electrical and Computer Engineering, University of California, Los Angeles, CA
3. Department of Chemical Engineering, Widener University, Chester, PA

With the evolution of modern-day electronics, the densification of transistors on semiconducting wafers has been accredited by the downscaling of transistor feature sizes and stacking of transistors, which give rise to high-performance technologies. However, the pursuit of these nanoscale devices has been challenging due to the stringent design criteria in the nanoscale and lack of control over the chemical processes required to assemble the transistors such that transistor surfaces are uniform to enable nanopatterning. To overcome this issue, thin-layer deposition techniques such as atomic layer deposition (ALD) have been implemented to deposit monolayers of film on the surfaces of the substrate in a self-limiting manner. Although this deposition process is effective in templating transistors, the procedure lacks controllability of the adsorption uniformity. Thus, a top-down approach using atomic layer etching (ALE) has been integrated in the semiconductor manufacturing process as a post-processing procedure succeeding ALD. Prior work [1] has established an *in silico* multiscale computational fluid dynamics (CFD) simulation that simulates an ALE process in a sheet-to-sheet (S2S) spatial reactor configuration as well as *ex situ* process modeling [2] to establish multivariate control to maintain thin-film conformity. However, this reactor model assumed that constant operating temperatures were maintained throughout the process to minimize the computational cost of simulating simultaneous momentum, heat, and mass differential equations spatiotemporally. Constant temperatures must be maintained due to the pyrolysis of precursor, trimethylaluminum, which decomposes at higher operating temperatures into various species [3] generated by changes in feed temperatures or dimerizes at lower operating temperatures [4] due to the observance of endothermic reactions. Additionally, high operating temperatures are required to ensure complete volatilization of byproducts to obtain an effective surface etch. Therefore, on-line temperature regulation is necessary to minimize the production of byproducts that may interfere with the self-limiting nature of the ALE process and control the fluid dynamics of the substrate exposure to the reagents [5].

This work proposes a CFD model conjoined with an on-line feedback control system to monitor the temperature operating conditions of the reactor, wafer surface, and the feed flow rate temperatures. This work will implement a methodical approach for conducting multivariate temperature adjustment in the presence of temperature disturbances such as ramp, step, and impulse disturbances. A model predictive controller designed on the basis of a data-based model is integrated into the CFD simulation via multiphysics simulation software, Ansys Fluent, through user-defined functions, and the controller performance and robustness properties will be evaluated through extensive closed-loop simulations under practical disturbances.

References:

1. Yun, S., Tom, M., Orkoulas, G., & Christofides, P. D., 2022. "Multiscale computational fluid dynamics modeling of spatial thermal atomic layer etching," *Computers & Chemical Engineering*, 163, 107861.
2. Tom, M., Yun, S., Wang, H., Ou, F., Orkoulas, G., & Christofides, P. D., 2022. "Machine learning-based run-to-run control of a spatial thermal atomic layer etching reactor," *Computers & Chemical Engineering*, 168, 108044, 2022.
3. Zhang, Z., Pan, Y., Jang, Z., & Fang, H., 2017. "Experimental study of trimethyl aluminum decomposition," *Journal of Crystal Growth*, 473, 6-10.
4. Adomaitis, R. A., 2018. "Estimating the thermochemical properties of trimethylaluminum for thin-film processing applications," *Journal of Vacuum Science & Technology A*, 36, 050602.
5. He, W., Chu, B., Gu, R., Zhang, H., Shan, B., & Chen, R., 2013. "The application of generalized predictive control in the radiant heating atomic layer deposition reactor," *2013 IEEE International Symposium on Assembly and Manufacturing (ISAM)*, Xi'an, 37-40.

135b Sample Size Determination for Metamodel Building in Automated Machine Learning Pipelines By an Inclusive Feedback Algorithm

Thalita C. R. Fernandes, Heleno Bispo, Antônio Carlos Brandão

This contribution aims at systematically determining the required sample size to build accurate metamodels by means of a novel framework that uses a feedback control strategy to drive an Automated Machine Learning pipeline, which is designed to produce simple regression models whenever possible, or else generate more complex, less desirable nonlinear models. At each iteration of the proposed algorithm, the feedback scheme computes the number of samples to be placed in the input space by sequential design of experiment techniques, which query the actual process on demand. The intended goal is to minimize the number of calls to the data generating system by processing all responses together within the same iteration step. This is achieved by a unique inclusive strategy where each generated dataset is exhaustively and effectively used up to converge even difficult responses until a stopping criterion on the error score is met. This contribution introduces three new features, which can be viewed as a paradigm shift in the field, namely: (i) a proportional controller to bring the error score to zero by determining the appropriate number of samples, (ii) a steady state detection routine to halt metamodel training when no progress is made, and (iii) an inclusive strategy to simultaneously handle multiple responses. The first two are brought from the field of Process Control Engineering, revealing the interdisciplinary nature of the algorithm. Application of the proposed framework to important test cases shows its effectiveness in building metamodels with significant predictive capabilities with the right number of samples.

Automated Machine Learning aims at compiling all the necessary decisions in a data-driven, objective, and automated way to produce an algorithm that makes the practice of model building more efficient for non-experts (Thornton et al., 2013). For this audience, there are many popular freely available AutoML algorithms, including Auto-WEKA (Kotthoff et al., 2016, 2019; Thornton et al., 2013), AutoGluon-Tabular (Erickson et al., 2020), Auto-sklearn (Feurer et al., 2015), H2O AutoML (LeDell & Poirier, 2020), Sumo (Gorissen et al., 2010), Hyperopt-sklearn (Komer et al., 2014), TPOT (Olson et al., 2016), Auto-Keras (Jin et al., 2019), to name a few. A high-level overview of the proposed heuristic inclusive feedback iterative AutoML pipeline (HIFIAPP) for regression of process data is shown in **Algorithm 1**. All these ideas were embedded in an Automated Machine Learning pipeline designed to systematically produce simple structured metamodels, using a minimum number of samples in an amenable amount of time. The application of the algorithm to representative test cases revealed the effectiveness of the proposed strategy to promote the construction of surrogate models with significant predictive capabilities.

References

Erickson, N., Mueller, J., Shirkov, A., Zhang, H., Larroy, P., Li, M., & Smola, A. (2020). *AutoGluon-Tabular: Robust and Accurate AutoML for Structured Data.*

Feurer, M., Klein, A., Eggensperger, K., Springenberg, J., Blum, M., & Hutter, F. (2015). Efficient and Robust Automated Machine Learning. In C. Cortes, N. Lawrence, D. Lee, M. Sugiyama, & R. Garnett (Eds.), *Advances in Neural Information Processing Systems 28 (NIPS 2015)* (Vol. 28). Curran Associates, Inc.

Gorissen, D., Couckuyt, I., Demeester, P., Dhaene, T., & Crombecq, K. (2010). A Surrogate Modeling and Adaptative Sampling Toolbox for Computer Based Design. *Journal of Machine Learning Research*, *11*, 2051–2055.

Jin, H., Song, Q., & Hu, X. (2019). Auto-Keras: An Efficient Neural Architecture Search System. *Proceedings of the 25th ACM SIGKDD International Conference on Knowledge Discovery & Data Mining*, 1946–1956. https://doi.org/10.1145/3292500.3330648

Komer, B., Bergstra, J., & Eliasmith, C. (2014). *Hyperopt-Sklearn: Automatic Hyperparameter*

Configuration for Scikit-Learn. 32–37. https://doi.org/10.25080/Majora-14bd3278-006

Kotthoff, L., Thornton, C., Hoos, H. H., Hutter, F., & Leyton-Brown, K. (2016). Auto-WEKA 2.0: Automatic model selection and hyperparameter optimization in WEKA. *Journal of Machine Learning Research*, *17*, 1–5.

Kotthoff, L., Thornton, C., Hoos, H. H., Hutter, F., & Leyton-Brown, K. (2019). *Auto-WEKA: Automatic Model Selection and Hyperparameter Optimization in WEKA* (pp. 81–95). https://doi.org/10.1007/978-3-030-05318-5_4

LeDell, E., & Poirier, S. (2020). H2O AutoML: Scalable Automatic Machine Learning. In F. Hutter, J. Vanschoren, M. Lindauer, C. Weill, K. Eggensperger, & M. Feurer (Eds.), *7th ICML Workshop on Automated Machine Learning*.

Olson, R. S., Bartley, N., Urbanowicz, R. J., & Moore, J. H. (2016). Evaluation of a Tree-based Pipeline Optimization Tool for Automating Data Science. *Proceedings of the Genetic and Evolutionary Computation Conference 2016*, 485–492. https://doi.org/10.1145/2908812.2908918

Thornton, C., Hutter, F., Hoos, H. H., & Leyton-Brown, K. (2013). Auto-WEKA. *Proceedings of the 19th ACM SIGKDD International Conference on Knowledge Discovery and Data Mining*, 847–855. https://doi.org/10.1145/2487575.2487629

Algorithm 1. A high-level overview of the proposed *heuristic inclusive feedback iterative AutoML pipeline (HIFIAPP)*.

```
Choose
     Case intersite-projected
     Case intersite-projected-threshold
     Case optimizer-projected
     Case optimizer-intersite
     Case lola-voronoi
End Choose
//Set parameters, or assume default values.
//Assign the regression method.
   If perctSamplesLinear = 0 Then
      Regression method for active responses ← iniModelNL.
      Regression methods for kresp ← algorithm chooses.
   Else
      Regression method for active responses ← iniModelLI.
      Regression methods for kresp ← set at the inner while-
      loop.
   End If
//Initialize variable for each response, and the active and
unconverged sets.
```

$m \leftarrow \max\left\{\left\lceil\frac{kfold}{kfold-1}\cdot \#inputs\right\rceil, 2\cdot kfold\right\}$ //initial points.

```
// Implement these initial points in the data generating
system.
```

// Compute RSE_m using these initial points.

Generate du ← 1 new point.

// Implement this single point in the data generating system and compute RSE_{m+du}.

$dRSE_{m+d} \leftarrow \frac{|RSE_{m+du}-RS\ \ m|}{du}$.

$RSE_m \leftarrow RSE_{m+d}$.

$m \leftarrow m + du$.

While (active ≠ Ø) **And** (time elapsed < maxTime) **Do**

//Generate du ← 2 new points.

//Implement these points in the data generating system and compute RSE_{m+du}.

$dRSE_{m+du} \leftarrow \frac{|RSE_{m+du}-RSE_m|}{du}$.

Calculate Q^2 for each response in the active and unconverged sets.

Select kresp as the response in the active set with the smallest Q^2.

samplesPerResp ← 0.

While (kresp exit criteria not met) **And** (samplesPerResp < maxSamplesPerResp) **And** (time elapsed < maxTime) **Do**

$RSE_m \leftarrow RSE_{m+d}$.

$m \leftarrow m + du$.

If samplesPerResp < maxSamplesLinear **Then**

Regression method for the kresp response ← iniModelLI.

Else

Regression methods for the kresp response ← ModelNL.

End If

$K_P \leftarrow \frac{|dRSE_{m+d}\ -dR\ \ m|}{du}$.

$K_C \leftarrow \begin{cases} \frac{1}{K_P}, & \frac{1}{K_P} \geq minK_C \\ minK_C, & otherwise \end{cases}$.

$du \leftarrow 0 + K_C(scoreCV_r(kresp) - 0)$

$du \in [1, maxDu]$.

// Generate du new points using the designated sequential design.

// Implement these points in the data generating system.

// Process responses in the background set and collect RSE_{m+d} .

// Process the kresp response by integer optimization and collect RSE_{m+d} .

$dRSE_{m+du} \leftarrow \frac{|RSE_{m+d} - RSE_m|}{du}$.

```
    For each response in the active set Do
        // Detect steady state on RSE (criterion 1).
```

```
        // Check whether the intercept of the RSE
    linearly regressed data (criterion 2).
        // Check whether all values of RSE are below
    user-defined threshold (criterion 3).
    End For
    If criterion 1 Or criterion 3 is satisfied, Then
      kresp ∈ converged set.
    End If
    If criterion 1 And criterion 2 are satisfied, Or
  criterion 3 is satisfied Then
      response ∈ converged set.
    End If
  samplesPerResp ← samplesPerResp + du.
End Inner While-Loop
If samplesPerResp > maxSamplesPerResp Then
  update the active, unconverged, and converged sets.
End If
```

$dRSE_m \leftarrow dRSE_{m+d}$.

```
End Outer While-Loop
Store relevant information.
```

135c Paraflow: Generative Design for Faster Additive Fabrication with Fewer Supports

Gabriel Lipkowitz, Ian Coates, Navneeth Krishna, Eric Shaqfeh, Joseph M. DeSimone

While 3D printing affords designers unprecedented geometric complexity in their products, currently the technology requires cumbersome support structures that are not user-friendly in numerous ways: materially wasteful, human labor-intensive, time-consuming to remove, damaging to surface finish, and often unreliable in ensuring printability in the first place. Inspired by recent physical demonstrations that injecting liquid through an embedded channel can neutralize suction during vat 3D printing, we discover this phenomenon can also significantly reduce the quantity of supports required by a user's model, or entirely obviate them for certain product geometries. To allow any resin 3D printer user to take advantage of this approach, we develop a novel fluid dynamics-guided computational inverse design tool, Paraflow, which innervates the user's arbitrary to-be-printed 3D model with a monotonic fluidic network for injection during printing. Our work draws on recent progress in generative artificial intelligence, in particular applied to search algorithms for 3D data trained on open source computer aided design (CAD) repositories such as Shapenet and Thingi10k, to optimize this network. Our goal is, for an arbitrary 3D model provided as input by a CAD modeler and 3D printer user, to construct a fluidic network that offsets suction, in order to effectively replace, or minimize, support structures required.

To generate such a suitable 3D printable network, we formulate the design problem as a path planning optimization problem where, for an arbitrary 3D geometry to be 3D printed via a set of 2D slice layers, we computationally design a corresponding fluidic network that innervates the part to sufficiently distribute one, or multiple, materials during printing. Our inverse design approach originates a fluidic network with a single input injection point connected to the build platform and, thus, the syringe pump. Then, injection channels are positioned within the user-provided model at optimal positions as determined by surrogate fluid dynamics modeling, in order to offset suction during printing to minimize need for supports. As it grows in complexity, the network is modeled as an electrical circuit, as is typical for pressure-driven fluidics. Branches may be either open (with outlets), or closed (without outlets), thus not contributing to fluid flow. The final network, encoded as a graph connectivity matrix of nodes and edges, is converted to a series of smooth NURBS curves, swept to produce a positive network 3D geometry, and finally Boolean differenced with the original B-rep CAD model via surface-to-surface intersection to produce negative channels, hence an innervated part. Rather than presenting the user with a single generatively designed fluidic network, Paraflow presents multiple options for innervating networks to the user, some of which may be more desirable for functional and/or aesthetic requirements. While there is one optimal network configuration that minimizes suction during printing, many possible configurations satisfy the design space constraints described above, with higher variance generally in part footprints with higher cross sectional area.

We experimentally demonstrate that Paraflow, applied to an off-the-shelf 3D printer with an inexpensive syringe pump add-on extension, enables designers to fabricate otherwise unprintable overhang geometries without support structures. When supports are not added to the model in the traditional process, print defects ranging from minor but noticeable to catastrophic result. With injection 3D printing and Paraflow computational design, however, such bridge overhangs are readily printable unsupported. To validate our fluidic inverse design approach, we describe our set-up for measuring forces during resin printing. In brief, we position both a load cell on the platform of our printer, to record suction forces, and a camera underneath the vat, to visualize the flow of injected resin. These online force sensors demonstrate that flow through these generatively-designed networks significantly reduce forces during printing, compared with control experiments without injection. This effect scales with the number of injection sites, as expected, and with the area of the printed object.

Finally, elastomers are among the most challenging materials with which to vat 3D print, due to aforementioned suction forces, but also among the most attractive for their energy absorbing or returning properties. We demonstrate that by overcoming suction with our generative design tool, we can print with highly flexible, shape-changing materials for applications in soft robotics and deformable interfaces, without support structures. Importantly, we characterize the limitations of our injection 3D printing generative design strategy, including the necessity to integrate hollow channels of non-zero thickness into parts, and describe post-curing strategies for mitigating this limitation. We also outline future directions for our novel generative AI design tool, including developing further software for fluid injection through multimaterial lattices, and integrating real-time computer vision control via neural networks trained on synthetically-generated data into injection 3D printers to allow users to better control multimaterial flows in real-time.

135d Implementing Statistical Process Control Using Python and Iiot for Real-Time Process Monitoring and Decision-Making

Gabriela Monteiro, Francisco Lucas Carneiro, Heleno Bispo

Statistical Process Control (SPC) is characterized by a set of statistical tools applied to production processes, allowing the systematic reduction of interference caused by the variability of special (non-natural) causes on the main quality characteristics of the product. If such problems are eliminated without the product being affected, the cost associated with reprocessing time and analysis of out-of-spec batches is reduced [1]. In the context of Industry 4.0 and the Industrial Internet of Things (IIoT), a variety of analysis and correlation technologies have emerged to predict situations based on data collected from multiple sources. Using a data infrastructure, it is possible to develop applications that utilize this information for process evaluation, operational, economic, or strategic decision-making, as well as for developing machine learning structures and phenomenological models with numerous applications [2].

Thus, the use of a data evaluation structure becomes paramount, and Python proves to be the ideal environment for such an implementation. The objective was to determine the Statistical Process Control (SPC) control limits using the stored data on the PI System (AVEVA/OSIsoft). The purpose of this approach was to evaluate an ongoing production process and facilitate comparative analysis of relevant literature. Once such limits determined, it is possible to be used on real-time analysis, as well as the development of structures for operational, economic and/or strategic decisions. To achieve this goal, a module was developed using Python and OPC communication to connect with the PI System database, storage, and visualization applications. The module was designed to create graphics according to the prescribed parameters [1; 3] for the control charts, enabling the construction of supervision screens that can be utilized for process monitoring and decision-making.

Keywords: SPC; Variability; IIoT; Data; Python.

References

[1] MONTGOMERY, D., Introduction to Statistical Quality Control, John Wiley and Sons, 2009.

[2] ROFFEL, B. E BETLEM, B., Process Dynamics and Control - Modeling for Control and Prediction, John Wiley & Sons, 2006.

[3] WHEELER, D., Advanced Topics in Statistical Process Control: The power of Shewharts Charts, SPC Press, 2004.

135e AI-Based Optimal Batch Control for Industrial Penicillin Fermentation Leveraging Deep Reinforcement Learning

Haoran Li, Tong Qiu

Starting with the development of deep tank fermentation in the 1940s, penicillin began to be produced on a large scale. As the decades progressed, penicillin fermentation has become a multi-billion-dollar industry. However, due to regulatory restrictions, the development of advanced process control technology for this process is still lagging compared to other chemical industries like oil and gas. As a pioneering batch biopharmaceutical process, kinetic simulations for penicillin fermentation have progressed. Currently, the most advanced industrial penicillin simulator is IndPenSim (http://www.industrialpenicillinsimulation.com). In this open-source simulator, the first-principles model of penicillin production is described by a system of ordinary differential equations incorporating a structured biomass model, component equilibria, and inhibition effects[1]. This simulator is also the benchmark for current research on advanced control strategies for penicillin fermentation processes[2].
The difficulty in batch control of the penicillin fermentation process comes from two main aspects. On the one hand, the process state measurements that can be observed during the process are limited. Measurements of the process are divided into three main areas: online measurements (cheap and fast), offline measurements (slow and with a lag), and Raman spectroscopy (expensive). Online measurements can provide a basic description of the process state such as temperature, pH, dissolved oxygen, fermenter weight, and outlet exhaust gas composition, but for some process critical variables such as penicillin concentration, biomass concentration, phenylacetic acid concentration, nitrogen concentration, and viscosity, despite their direct and significant effect on the penicillin growth rate, they can only be obtained by offline measurements, which provide only lagging, low-resolution data, making it difficult to construct high-precision control strategies. In addition, Raman spectroscopy-based process analytical technique (PAT) can predict phenylacetic acid concentration in real-time based on spectral data, thus enabling real-time PID control of phenylacetic acid concentration, but this technique is expensive, and not many practical applications are available.
On the other hand, process disturbances make the batch yield of penicillin unstable. For batch production, there are differences in the initial fermentation conditions of each batch, such as initial substrate concentration, the specific growth rate of biomass, dissolved oxygen, etc., and the growth rate of penicillin is very sensitive to these variables, so when the initial conditions are different, the batch yield can vary significantly from under the same control strategy. In addition, there will be in-batch fluctuations during each batch production. Such as inlet sugar concentration, oil concentration, phenylacetic acid concentration, cooling water temperature, etc., these exogenous disturbances can also disturb the state inside the fermenter and make the batch yield fluctuate.
Due to these two difficulties, the three control strategies reported so far: operator control, recipe-driven control, and PAT control are not good enough to achieve optimal batch control under disturbances with limited observation space. In recent years, with the development of deep reinforcement learning (DRL), great successes have been achieved in the fields of gaming, aerospace, and autonomous driving. Reinforcement learning uses multilayer neural networks to approximate value and policy functions, and can automatically find the action that maximizes the desired future reward in a given state by learning from interaction with the environment, thus achieving optimal control of the process. However, the application of this technique in the field of chemical process control is still limited.
In this work, we use DRL to achieve optimal batch control of the penicillin fermentation process. First, three elements in reinforcement learning are defined: state, action, and reward. The state is a 16-dimensional vector describing the penicillin fermenter, containing five offline measurement variables with a sampling frequency of 12 hours and 11 online measurement variables with a sampling frequency of 12 minutes. Action is a 7-dimensional vector containing all manually controlled variables of the process. The return is the amount of change in the yield of penicillin in the fermenter at that moment compared to the previous moment, obtained by subtracting the yield of the previous step using the yield of the current step. The time interval for each step in the training process is 12 minutes. The interaction between the agent and environment goes in the following way: In each step, the agent receives state variables from the IndPenSim

environment (PenSimEnv) and then gives out the action. The action is sent to PenSimEnv for execution. By solving the ODEs, PenSimEnv will generate the current step reward and the next state, which are used for the agent updates.

In this study, the Soft Actor-Critic (SAC) algorithm[3] was employed to train our model. This algorithm has gained significant attention in recent years for its effectiveness in solving complex continuous control problems. Unlike traditional Q-learning approaches, SAC is a policy optimization method that directly learns a stochastic policy in an off-policy setting. This unique feature makes SAC suitable for handling both discrete and continuous action spaces. SAC employs maximum entropy reinforcement learning to encourage exploration and a critic network to estimate the value function. In addition, it employs a soft Q-learning objective to learn a more effective policy. The combination of these techniques allows SAC to learn efficient policies for various challenging robotic and control tasks.

The training episode is set as 10000, and the steps in each episode are 1150. After the training, 30 test batches are used for comparing the performances of the developed DRL model with three reported control strategies (operator, recipe-driven, and Raman-spectroscopy-based PAT). The average batch yields for operator, recipe-driven, PAT, and DRL models are 2552, 2879, 3517, and 3810 kg. In addition, the DRL model demonstrated the smallest yield variance among the 30 batches of yields. The above results indicate that the DRL model can significantly improve the batch average yield while reducing the fluctuation of the batch yield, demonstrating its superiority in comparison with existing methods.

In summary, we present an advanced DRL-based optimal batch control framework for industrial penicillin production. The DRL framework enables implicit learning of the process state using limited offline and online observations to achieve higher batch yields and smaller yield fluctuations than existing control methods in the presence of process disturbances. We believe that the proposed framework is a promising artificial intelligence-based batch control method for industrial processes.

(1) Goldrick, S.; Ştefan, A.; Lovett, D.; Montague, G.; Lennox, B. The Development of an Industrial-Scale Fed-Batch Fermentation Simulation. Journal of Biotechnology 2015, 193, 70–82. https://doi.org/10.1016/j.jbiotec.2014.10.029.

(2) Goldrick, S.; Duran-Villalobos, C. A.; Jankauskas, K.; Lovett, D.; Farid, S. S.; Lennox, B. Modern Day Monitoring and Control Challenges Outlined on an Industrial-Scale Benchmark Fermentation Process. Computers & Chemical Engineering 2019, 130, 106471. https://doi.org/10.1016/j.compchemeng.2019.05.037.

(3) Haarnoja, T.; Zhou, A.; Abbeel, P.; Levine, S. Soft Actor-Critic: Off-Policy Maximum Entropy Deep Reinforcement Learning with a Stochastic Actor. 2018. https://doi.org/10.48550/ARXIV.1801.01290.

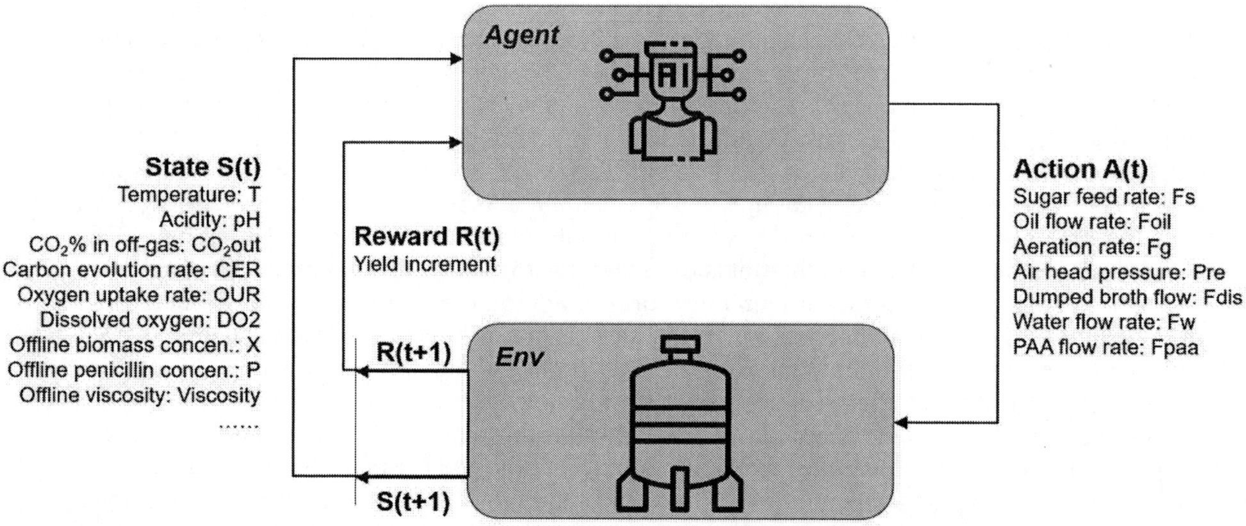

135f Applications of Artificial Intelligence for Thermal Analysis of Heat Exchangers

Emmanuel Dada, Mario Lumueno, Elizabeth Osadare, Sarhan Musa

Artificial Intelligence (AI) is gaining prominence as a technology that can be used to properly perform thermal analysis of heat exchangers with acceptable precision. In the last two decades, AI has been widely employed for the thermal analysis of heat exchangers. The applications of artificial intelligence (AI) on heat exchangers are discussed in this paper. The published investigations on the heat exchanger and thermal analysis are divided into four categories: heat exchanger modeling, estimation of heat exchanger parameters, estimation of phase change characteristics in heat exchangers, and heat exchanger control. This paper focuses on the two methods to apply artificial intelligence on heat exchangers which are the Artificial Neural Networks (ANN) and the Neuro-Fuzzy control also known as the Fuzzy Logic Controller (FLC). ANN is an information processing paradigm based on the human brain's biological neuron system. The limits of ANN for thermal analysis of heat exchangers are discussed, as well as future research needs in this sector. A fuzzy logic control model was integrated with neural network approaches. MATLAB software is used to simulate fuzzy logic controller models. Artificial Neural Network (ANN) models for simulation, optimization, and performance prediction of thermal systems incorporating heat exchangers were developed to tackle this challenge. The findings reveal that the fuzzy logic controller's control is capable of effectively stabilizing the temperature of the heat exchangers. The approach is more precise since the ANN comprises a large number of processing units in the network that execute concurrent data processing.

196a Keynote Talk: Cybersecurity and High-Performance Computing

Sandra K. Parker, Todd Fox, Byron Huber, Mike Fazio

As the world, and especially those in the chemical industry, progress forward with sustainability and digitalization goals, there is an increasing need to leverage High Performance Computing (HPC) capabilities to drive innovation and solve complex problems. When we think about HPC solutions that we deploy in our environment, finding a balance between information sharing and protection of intellectual property is challenging but is manageable by implementing security by design concepts to ensure 'cyber' is addressed from the inception and integrated fully with design, testing and implementation.

High-performance Computing is a technology that uses supercomputer and computer clusters to solve advanced computation problems. These systems typically perform at speeds more than one million times faster than the fastest desktop or laptop systems.[1] High-performance computing serves as a catalyst for understanding what the next generation will look like. No longer is this work done with on premise devices and servers, but we are progressing more to work in a cloud-enabled environment and sharing data with external parties that are involved in joint research projects. Transporting data allows for more rapid access to the data however this also requires us to be more diligent when it comes to data exposure and properly protecting the intellectual property associated with it.

Within companies like Dow, we leverage these devices to assist with aggregating data built on the need for computational development of understanding dynamics and general chemistry associated with the ability to determine impact with new ideas, areas we are trying to solve or the effects of change itself.

Collaborating with our businesses is critical in understanding what risk will be tolerated in this environment to drive what type of cybersecurity controls must be in place to minimize any data loss and provide visibility who is accessing this data. High Performance computing in an era where cybersecurity attacks are prevalent places even more emphasis on Cybersecurity design principles with regards to vulnerability and perimeter management. Some of the learnings and cybersecurity best practices will be discussed during the presentation.

References

[1] Wikipedia (High-performance computing - Wikipedia)

196b Keynote Talk: Cyber-Secure Machine Learning Modeling and Predictive Control of Nonlinear Chemical Process Network Using Federated Learning

Zeyuan Xu, Zhe Wu

The security of process control systems has become crucially important since control systems are vulnerable to cyber-attacks, which are a series of computer actions employed by the attacker to compromise the security of control systems. As machine learning (ML) approaches have been widely used to develop data-driven models for nonlinear systems [1], cybersecurity and data privacy have garnered increasing research interests with the use of big data. For example, developing an ML-based detector for large-scale distributed systems requires a tremendous volume of data to be collected from all subsystems through various mediums of communication such as Internet and wireless networks, and then processed in a central server or cloud for training. However, as communication mediums are vulnerable to attackers, the ML-based detector developed on a local server or cloud could be misguided and unable to detect target cyber-attacks in the presence of data tampering or data manipulation [2, 3]. Therefore, further advances in techniques and frameworks for promoting data privacy are needed to provide tractable solutions for the implementation of ML modeling methods.

To alleviate the security concerns, a federated-learning-based model predictive control (FL-based MPC) method for nonlinear chemical process network is developed in this work. By taking advantage of the idea of federated learning that distributes a pre-trained model to all subsystems and allows each subsystem to develop and update its own model locally without sharing the raw data with the central server, we first develop an FL framework with fully-connected and partially-connected neural network structures to model the entire process network with enhanced data privacy while accounting for the heterogeneity of nonlinear systems with multiple subsystems (note that the heterogeneity of the subsystems will result in non-independent and identically distributed training data [5]). Subsequently, we develop a theoretical generalization error bound and the estimation of privacy leakage for the FL models using relative entropy and duality theory. The closed-loop stability of distributed systems under the FL-based MPC approach is further developed. Finally, a chemical reactor is used as an example to demonstrate the effectiveness of the proposed FL modeling and FL-based MPC approach.

[1] Wu, Z., Tran, A., Rincon, D., & Christofides, P. D. (2019). Machine learning-based predictive control of nonlinear processes. Part I: theory. AIChE Journal, 65(11), e16729.

[2] Tan, A. Z., Yu, H., Cui L., & Yang, Q. (2022). Towards Personalized Federated Learning. IEEE Transactions on Neural Networks and Learning Systems, doi: 10.1109/TNNLS.2022.3160699.

[3] Chen, M., Yang, Z., Saad, W., Yin, C., Poor, H. V., Cui, S. (2021). A Joint Learning and Communications Framework for Federated Learning Over Wireless Networks. IEEE Transactions on Wireless Communications, 20(1), 269-283.

[4] Christofides, P. D., Scattolini, R., Pena, D. M., & Liu, J. (2013). Distributed model predictive control: A tutorial review and future research directions. Computers and Chemical Engineering, (51), 21-41.

[5] Sattler, F., Wiedemann, S., Müller, K. R., & Samek, W. (2020). Robust and Communication-Efficient Federated Learning From Non-i.i.d. Data. IEEE Transactions on Neural Networks and Learning Systems, 31(9), 3400-3413.

196c Control Mode Selection for Switching-Enabled Cyberattack Detection

Shilpa Narasimhan, Nael El-Farra, Matthew Ellis

Process control systems (PCSs) are industrial control systems that operate chemical manufacturing processes and have been targets for several sophisticated cyberattacks [1], leading to a focus on enhancing their cybersecurity [2]. PCS designs for cybersecurity consider the scenario in which a malicious agent can alter operational data or control logic after successfully circumventing information technology-based cybersecurity protections (e.g., firewalls). Several approaches have been presented for cybersecure PCS design [3]-[11], including attack detection schemes to determine the presence of an attack (e.g., [3], [9]-[11]) and attack identification and mitigation strategies (e.g., [7]). A cyberattack detection scheme detects an attack on the process based on anomalous behavior of measured variables.

Control parameters may influence the ability of a detection scheme to detect an attack [8]. The control mode that a process is operated under for extended periods may be designed to use control parameters that ensure that the closed-loop performance requirements are met. Under this mode, called the nominal control mode, some attacks may not be detectable, and for their detection, the process may need to be operated under alternative control modes. Alternative control modes may be selected to ensure attack detection and may not meet the closed-loop performance requirements for the process. To manage a potential tradeoff between attack detection and performance requirements, the control mode may be switched occasionally to an alternative control mode. Control mode switching may excite process dynamics and induce a transient behavior in the process operated under steady-state conditions (when the process states evolve within a neighborhood of the steady-state and the monitoring variable evolves within a terminal set). Transient behavior may trigger undesirable false alarms in a terminal set-based detection scheme [9], [10]. Since reachable sets of the attack-free process track the evolution of the monitoring variable, a reachable set-based attack detection scheme [11] is utilized to monitor the switched process for attacks while ensuring zero false alarms from switching. An attack is detected if the monitoring variable is not contained within the attack-free reachable set.

This work presents strategies for selecting alternative control modes for switching-enabled cyberattack detection using the reachable set-based detection scheme. Several strategies are presented and compared. The first strategy ensures that the terminal set of the monitoring variable under the alternative mode is larger than the terminal set under the nominal control mode. The second strategy is to use an alternative control mode under which an attack on the process destabilizes it [9], [10]. Finally, the third strategy involves selecting control parameters chosen such that the reachable sets of the attacked and the attack-free processes do not intersect at some time step, enabling attack detection. To ensure the detection of a wide range of attacks, different control modes are used over the length of the operation. Using an illustrative process example, the selection of and implementation of each alternative control mode to enable the detection of attacks with zero false alarms is demonstrated.

References:

[1] T. Miller, A. Staves, S. Maesschalck, M. Sturdee, and B. Green, "Looking back to look forward: Lessons learnt from cyber-attacks on industrial control systems," International Journal of Critical Infrastructure Protection, volume 35, pp. 100464, 2021.

[2] M. S. Mahmoud, M. M. Hamdan, and U. A. Baroudi, "Modeling and control of Cyber-Physical Systems subject to cyber attacks: A survey of recent advances and challenges," Neurocomputing,

volume 338, pp. 101-115, 2019.

[3] D. Zhang, Q. G. Wang, G. Feng, Y. Shi, and A. V. Vasilakos, " A survey on attack detection, estimation and control of industrial cyber–physical systems," ISA transactions, vol. 116, pp. 1-16, 2021.

[5] S. Parker, Z. Wu, and P. D. Christofides, "Cybersecurity in process control, operations, and supply chain," Computers & Chemical Engineering, volume 171, pp. 108169, 2023.

[6] K. Neiman, D. Messina, M. Wegener, and H. Durand, "Cybersecurity and dynamic operation in practice: Equipment impacts and safety guarantees," Journal of Loss Prevention in the Process Industries, volume 81, pp. 104898, 2023.

[7] A. Zedan, and N. H. El-Farra, "A machine-learning approach for identification and mitigation of cyberattacks in networked process control systems," Chemical Engineering Research and Design, volume 176, pp. 102-115, 2021.

[8] S. Narasimhan, N. H. El-Farra, and M. J. Ellis, "Detectability-based controller design screening for processes under multiplicative cyberattacks," AIChE Journal, volume 68, pp. e17430, 2022.

[9] S. Narasimhan, N. H. El-Farra, and M. J. Ellis. "Active multiplicative cyberattack detection utilizing controller switching for process systems," Journal of Process Control, volume 116, pp. 64-72, 2022.

[10] S. Narasimhan, N. H. El-Farra, and M. J. Ellis. "A control-switching approach for cyberattack detection in process systems with minimal false alarms," AIChE Journal, volume 68, pp. e17875, 2022.

[11] S. Narasimhan, N. H. El-Farra, and M. J. Ellis. "A reachable set-based cyberattack detection scheme for dynamic processes," Proceedings of the American Control Conference, In Press, 2023.

196d Fighting Back Against Cyberattacks on Control Systems Under Lyapunov-Based Economic Model Predictive Control

Dominc Messina, Keshav Kasturi Rangan, Helen Durand

Cyberattacks on cyber-physical systems (CPSs) may affect components of a control hierarchy, such as sensors and actuators. Different challenges arise in detecting and fighting back against cyberattacks on different components which must be accounted for, or there becomes a potential for loss of process control and damage to equipment and the surroundings. Previous work in our group included development of detection strategies using LEMPC to flag cyberattacks on a system. Detection of an attack, depending on the strategy, may guarantee process safety for some time after an attack occurs; however, the guarantees of safety in, for example, [1,2] end at the latest after attacks are detected. In [3], for example, it was assumed that cyberattacks could be detected and that then back-up sensors could be applied to re-gain safety. However, such a concept would require adequate redundancy and diagnosis. The means for achieving adequate diagnosis and trying to reduce redundancy with the control-theoretic framework based on Lyapunov-based economic model predictive control (LEMPC) [4] developed in our prior work for detecting cyberattacks requires further investigation.

In this talk, we will discuss ideas for extending our prior detection strategies that integrate with LEMPC to diagnosis and post-detection handling methods. Specifically, we will initially review results with regard to attempts to use distributed controllers for sensor attack diagnosis in [2]. We will also consider other ideas for attempting to diagnose attacks. For example, one of the detection strategies of our prior work relies on randomly modifying the LEMPC used for controlling the process such that it drives the closed-loop state toward a lower level set around some new steady-state. This helps to showcase an attack if the Lyapunov function does not decrease. We will use an example of a continuous stirred tank reactor (CSTR) to evaluate whether the different terms of the time derivative of a Lyapunov function, and their degree of negativity, could aid with diagnosing attacks on actuators. We will also consider methods for updating the constraints of an LEMPC after an attack is detected and diagnosed that assume a variety of potential behavior from the attacker and seek to maintain safety regardless (or in as many potential situations as possible). We will discuss ways of formulating this problem, and the extent to which guarantees can be made.

After these preliminary studies in investigating ideas for using the LEMPC to 'fight back' against the attacker through diagnosis and subsequently incorporating knowledge of the attack in the control action selection, we will consider the problem of how to obtain the original control-theoretic guarantees before detection. LEMPC is a control strategy which has been developed for the control of nonlinear systems. LEMPC is a particularly attractive control strategy due to its strong theoretical guarantees regarding stability and feasibility, however, there does not exist a standard method to develop all of parameters of this controller needed for any process to meet these theoretical guarantees. In our prior work, we have discussed preliminary ideas for attempting to obtain these parameters, including a 'best-case' analysis based on assuming certain values for the parameters and the absence of disturbances, and then solving for the largest possible sampling period that would meet the resulting constraints [2]. We also considered an optimization problem [5] that attempts to find parameters that cause the control-theoretic guarantees to be met at some points in the state-space. These initial results were investigated in the context of the standard LEMPC from [4]; however, the attempts to obtain the parameters may impact not only the practical implementation of LEMPC itself, but also control strategies based on LEMPC for applications such as cybersecurity. For the guarantees of safety related to the pre-detection period from our prior work to be obtained, theoretical conditions related to LEMPC must be again met in the design stage of the controller and during operation. In this final part of the talk, we will extend the methods from [2,5] towards the development of a cyberattack detection strategy based on LEMPC toward a CSTR example to characterize conditions which would make the sufficient conditions for cyberattack detection and safety guarantees related to the detection procedures hold. We will discuss the limitations related to the practical implementation of such a strategy in regards to the design of a controller and the conditions under which detection can be guaranteed.

[1] Oyama, Henrique, and Helen Durand. 'Integrated cyberattack detection and resilient control strategies using Lyapunov-based economic model predictive control.' *AIChE Journal* 66.12 (2020): e17084.

[2] Oyama, Henrique, et al. 'Lyapunov-based economic model predictive control for detecting and handling actuator and simultaneous sensor/actuator cyberattacks on process control systems.' *Frontiers in Chemical Engineering* 4 (2022): 810129.

[3] Wu, Z., Albalawi, F., Zhang, J., Zhang, Z., Durand, H., & Christofides, P. D. (2018). Detecting and handling cyber-attacks in model predictive control of chemical processes. *Mathematics*, *6*(10), 173.

[4] Heidarinejad, M., Liu, J., & Christofides, P. D. (2012). Economic model predictive control of nonlinear process systems using Lyapunov techniques. *AIChE Journal*, *58*(3), 855-870.

[5] Nieman, K., Messina, D., Wegener, M., & Durand, H. (2023). Cybersecurity and dynamic operation in practice: Equipment impacts and safety guarantees. *Journal of Loss Prevention in the Process Industries*, *81*, 104898.

196e On Robustness of Encrypted Model Predictive Control of Nonlinear Processes

Atharva Vijay Suryavanshi, Aisha Alnajdi, Fahim Abdullah, Panagiotis Christofides

With the increased risk of targeted cyber-attacks, it has become essential to incorporate cyber-secure approaches into physical networked control systems to ensure safe and secure operations. The use of unsecure networks to facilitate communication between various components of networked control systems, along with computations involving sensitive data on outsourced platforms, can result in the risk of data manipulation and interception. Several approaches have been used to ensure cyber-security in networked control systems, some of which include machine-learning based detectors for advanced threat detection [1], design of two-tier controllers [2] and encrypted control systems [3]. In our previous work [4], we developed a Lyapunov-based encrypted Model Predictive Control (MPC) scheme for nonlinear systems, which involved establishing secure communication in the sensor-controller and controller-actuator links. Due to the nature of cryptosystems, real valued signals have to be mapped to a set of positive integers, on which the encryption-decryption operations are carried out. This mapping results in quantization errors, and, in our previous work, we analyzed the effect of these quantization errors on closed-loop stability of the nonlinear system.

In this work, we use the previously developed encrypted Lyapunov-based MPC scheme to control a chemical process simulated using the well-known high-fidelity process simulator, Aspen Plus Dynamics. As the model predictive controller uses a first-principles model of the process, there is a model mismatch between the model used in the controller and the Aspen model. Specifically, we analyze the combined and relative effects of the quantization errors, arising due to the encryption operations, and the error due to the plant-model mismatch on closed-loop stability of the nonlinear system.

References:

[1] Omar S, Ngadi A, Jebur HH. Machine learning techniques for anomaly detection: an overview. *International Journal of Computer Applications*. 2013;79:33–41.

[2] Chen S, Wu Z, Christofides PD. A cyber-secure control-detector architecture for nonlinear processes. *AIChE Journal*. 2020;66:e16907.

[3] Darup MS, Redder A, Shames I, Farokhi F, Quevedo D. Towards encrypted MPC for linear constrained systems. *IEEE Control Systems Letters*. 2017;2:195–200.

[4] Suryavanshi A., Alnajdi A, Alhajeri M, Abdullah F, Christofides PD (in press). Encrypted Model Predictive Control Design for Security to Cyber-Attacks. *AIChE Journal*.

196f Handling Cyberattacks in Advanced Control of Cubesats

Jihan Abou Halloun, Keshav Kasturi Rangan, Helen Durand

Satellites are critical to day-to-day life, with their applications ranging from communication to observations helping scientists to learn about Earth and space. Cyberattacks on satellites are typically well managed in larger programs like the Joint Polar Satellite System and Orion [6], but smaller low-cost missions like CubeSats, a class of nanosatellites, could be vulnerable to attack such as spoofing attacks [7]. CubeSats are a 4-inch cube that are characterized by their light weight. They have been employed in, for example, interplanetary missions [8]. Cyberattacks such as spoofing can be problematic on satellites that incorporate control strategies which manipulate actuators leading to modification of their orientation and degrading their performance. Cyberattacks on satellite have been discussed in the literature. [1] for example, discusses a potential outcome of a cyberattacks involving satellites as their ability to harm one another. As a step toward cyberattack-resilient control of autonomous cyber-physical systems, [2] proposed a resilient estimator where Kalman filtering technique combined with watermarking approach was elaborated in order to detect cyberattacks on a CubeSat's sensors such as spoofing. A number of strategies for handling cyberattacks on control systems were also developed by our group in order to detect and mitigate cyberattacks for systems described by nonlinear differential equations (e.g., [3][4][5]). However, the needs of space systems (e.g., reduced computation time) were not directly addressed in our group's prior work.

In this work, we analyze how our prior methods extend to the case of the CubeSat control. We begin by discussing the dynamic model used to represent the CubeSat behavior. We then discuss the theoretical and application principles of applying our prior cyberattack detection policies that integrate with an advanced control algorithm known as Lyapunov-based economic model predictive control (LEMPC) [9], which were developed in [10] to the case of the CubeSat. We then discuss how the strategies in [10] could be extended to allow other advanced controllers that are not LEMPC to be considered with respect to their cybersecurity guarantees. For example, we consider a case of an EMPC with a terminal constraint instead of an LEMPC, and how it might be integrated with detection strategies. We compare the different strategies developed with respect to their effectiveness and ease of use to provide a more comprehensive view of different integrated cyberattack detection and control policies than have been developed in our prior work.

References:

[1] Falco, G. (2020). When Satellites Attack: Satellite-to-Satellite Cyber Attack, Defense and Resilience. In *ASCEND 2020* (p. 4014).

[2] Marquis, V., Ho, R., Rainey, W., Kimpel, M., Ghiorzi, J., Cricchi, W. and Bezzo, N., 2018, April. Toward attack-resilient state estimation and control of autonomous cyber-physical systems. In *2018 Systems and Information Engineering Design Symposium (SIEDS)* (pp. 70-75). IEEE.

[3] Helen Durand. "A nonlinear systems framework for cyberattack prevention for chemical process control systems". In: Mathematics 6.9 (2018), p. 169.

[4] Helen Durand and Matthew Wegener. "Mitigating cyberattack impacts using Lyapunov-based economic model predictive control". In: 2020 American Control Conference (ACC). IEEE. 2020, pp. 1894–1899.

[5] Keshav Kasturi Rangan, Henrique Oyama, and Helen Durand. "Integrated cyberattack detection

and handling for nonlinear systems with evolving process dynamics under Lyapunov-based economic model predictive control”. In: Chemical Engineering Research and Design 170 (2021), pp. 147–179.

[6] NASA Office of Inspector General, Office of Audits, NASA’s Cybersecurity Readiness, Report No. IG-21-019, May 18, 2021

[7] Marquis, V., Ho, R., Rainey, W., Kimpel, M., Ghiorzi, J., Cricchi, W., & Bezzo, N. (2018, April). Toward attack-resilient state estimation and control of autonomous cyber-physical systems. In *2018 Systems and Information Engineering Design Symposium (SIEDS)* (pp. 70-75). IEEE.

[8] Howell, Elizabeth. 'Cubesats: Tiny payloads, huge benefits for space research.' *Space. com* (2018).

[9] Heidarinejad, M., Liu, J., & Christofides, P. D. (2012). Economic model predictive control of nonlinear process systems using Lyapunov techniques. *AIChE Journal*, *58*(3), 855-870.

[10] Oyama, H., & Durand, H. (2020). Integrated cyberattack detection and resilient control strategies using Lyapunov-based economic model predictive control. *AIChE Journal*, *66*(12), e17084.

196g Cyberattack-Resilient Data-Gathering Policies Under Lyapunov-Based Economic Model Predictive Control for Physics-Based Model-Building

Akkarakaran Leonard, Henrique Oyama, Gabriele Rabajoli, Helen Durand

Data-gathering maneuvers by controllers have been considered in the context of, for example, dual model predictive control [4]. Attempts to use data in the identification of physics-based models have been pursued in, for example, [5,6]. However, it remains challenging to identify physics-based models from data. In [2,3], we attempted to use a controller to perform data-gathering maneuvers to identify physics-based models. However, an open question remains how to tell which data to gather to ensure that it identifies a physics-based model. On a different note, data has potential to be compromised if sensors are compromised. One of the main contributions of our recent work in [1] is that it elucidates how to detect attacks on an advanced controller known as Lyapunov-based economic model predictive control (LEMPC) [7] when sensor attacks,actuator attacks, or both can occur.

In this work, we consider two questions: 1) how to identify what data should be gathered for developing physics-based models and 2) how to extend our prior concepts on data-gathering to handling whether the data is correct or incorrect (both before the model is re-identified through detection and through analyzing the impacts of it being incorrect when the model is re-identified). We recall from [2,3] how LEMPC can be utilized for online data collection once desired data is identified. We then discuss methods for attempting to identify that data, including making use of concepts from Automated Theorem Proving. Specifically, Automated Theorem Proving (ATP) is a method that has seen applications in mathematical fields to automatically prove conjectures within a logical framework. This framework is formed from logical statements known as axioms, which are considered to be always true. In order to utilize the principles of ATP to aid in model validation, a method is proposed to attempt to verify whether a model is correct according to whether it meets certain axioms that should be satisfied by the data. We analyze this concept initially with a level in a tank process for a pre-developed axiom specifying that a correct model for the process should meet a certain requirement related to increases and decreases in height. We then explore how axioms might be automatically developed. After methods for identifying desired data are identified, we discuss how to use LEMPC for driving the closed-loop state toward that data as in [2,3], but add additional discussion regarding checking for cyberattacks during the data-gathering phase according to methods like those developed in [1,8]. We then also discuss how un-detected attacks during the data-gathering stage could impact model re-identification, considering a control-theoretic viewpoint in an LEMPC context integrated with the various detection strategies.

[1] Oyama, H., Messina, D., Durand, H., Rangan, K., "Lyapunov-Based Economic Model Predictive Control for Detecting and Handling Actuator and Simultaneous Sensor/Actuator Cyberattacks on Process Control Systems," Frontiers in Chemical Engineering, 2022; 810129.

[2] Oyama, H., Durand, H., "Lyapunov-Based Economic Model Predictive Control for Online Model Discrimination," Computers Chemical Engineering, 2022; 107769.

[3] Oyama, H., Leonard, A. F., Rahman, M., Gjonaj, G.,Williamson, M., and Durand, H., "Online Process Physics Tests via Lyapunov-based Economic Model
Predictive Control and Simulation-Based Testing of Image-Based Process Control," American Control Conference, paper 1187, 2021.

[4] Heirung, T. A. N., Ydstie, B. E., & Foss, B. (2012). Towards dual MPC. *IFAC Proceedings*

Volumes, *45*(17), 502-507.

[5] Brunton, S. L., Proctor, J. L., & Kutz, J. N. (2016). Discovering governing equations from data by sparse identification of nonlinear dynamical systems. *Proceedings of the national academy of sciences*, *113*(15), 3932-3937.

[6] Schmidt, M., & Lipson, H. (2009). Distilling free-form natural laws from experimental data. *Science*, *324*(5923), 81-85.

[7] Heidarinejad, M., Liu, J., & Christofides, P. D. (2012). Economic model predictive control of nonlinear process systems using Lyapunov techniques. *AIChE Journal*, *58*(3), 855-870.

[8] Oyama, H., & Durand, H. (2020). Integrated cyberattack detection and resilient control strategies using Lyapunov-based economic model predictive control. *AIChE Journal*, *66*(12), e17084.

268a First Steps Toward Developing an Image-Controlled Material Assembly

Dominc Messina, Helen Durand

In recent years, the rise of digital sensing modalities such as image sensors has opened opportunities for the development of advanced control laws utilizing this data. One such application is known as image-based predictive control, referring to the concept of using predictions of how a scene will look in the future to indicate predicted process states to be used by a controller to determine optimal inputs. Image-based predictive control has seen applications in the areas such as visual servoing [3] and autonomous vehicles [1], in which either mathematical transformations or artificial intelligence [2] are used to predict future states of the system. Image-based control has potential to be used in a variety of applications. One potential future application could be with respect to control of materials. Controlled material assemblies is a concept for next-generation materials, referring to the use of control to modify properties or the structure of an assembly. One might imagine images as a key sensing mechanism in control of materials.

This talk focuses on the development of a framework for creating a 'mock-up' of an image-controlled material that can be used as a first step toward attempting to design a controlled material with desired properties that also responds to images. In general, the concept of developing a material that responds to images in a desired fashion requires a number of components, including: 1) identifying the material behavior desired (and then how to achieve it); 2) developing the controller's working specifications; and 3) developing the appropriate algorithm for image-based control/processing. We would like to break this problem into pieces, so that first the major functionality of the material is assessed and designed, so that it can then be considered how and whether a material meeting the resulting requirements can be designed.

The simulation will consider that we would like to design a material with the specification that it should break when certain visuals are created. We would like it to have this behavior progressively (i.e., to predict that it might see the problematic visual that would cause it to break and to start to prepare for this so that it does not have a large lag for breaking when the problematic visual is observed). To handle the importance of the image components in this simulation, we present the development of the control algorithm and image-handling methods in the computer graphics software Blender in which a mock-up material assembly made of cubes that are considered to be able to be ejected from the assembly toward a new position at a rate given by a first-order dynamic model is discussed. A camera is placed in the Blender scene to represent the image sensor. It is pointed at a different cube, which is the one where, when its vertices exhibit certain colors at a certain position, signifies the problematic condition that should cause the material to break. We will use an image prediction algorithm to predict where the vertices might end up based on their current positions as rendered by the camera in Blender, and different possible concepts for what the cube might look like that could result in different future visuals. The image prediction algorithm will build on our prior preliminary results using OpenGL [4] for coding different visuals of a cube. The goal of the image prediction algorithm is to modify the camera position to reveal the nature of the block visuals. We will discuss different potential formulations of the image prediction-based control algorithm (e.g., different objective functions) and their effect on the movement of the camera and what it sees. The predictions will be made using image transformations relate to those which would be used in OpenGL for image rendering, but coded in Blender's Python programming interface to avoid the render time. The vertices of the cube in the camera's field of view will be captured using a render in Blender that will then be processed to provide the vertex information. The transformations of the vertices in the image prediction algorithm will be according to the predicted movement of the cube to yield predictions of what the images in the camera's field of view will look like at discrete times in the future. In this example, each face of the cube is a different color, and the cube rotates continuously. We will demonstrate how the controller ejects cubes from the material for different algorithm formulations and cube visuals, to better understand the responses of the material and tuning of both its behavior and the control and image prediction algorithms for this application. This initial mock-up will then be used to suggest what types of properties might be desirable in a material that could have this type of responsiveness.

[1] Lee, Daewon, Hyon Lim, and H. Jin Kim. 'Obstacle avoidance using image-based visual servoing integrated with nonlinear model predictive control.' 2011 50th IEEE Conference on Decision and Control and European Control Conference. IEEE, 2011.

[2] Oprea, Sergiu, et al. 'A review on deep learning techniques for video prediction.' IEEE Transactions on Pattern Analysis and Machine Intelligence 44.6 (2020): 2806-2826.

[3] Sheng, Huaiyuan, Eric Shi, and Kunwu Zhang. 'Image-based visual servoing of a quadrotor with improved visibility using model predictive control.' 2019 IEEE 28th international symposium on industrial electronics (ISIE). IEEE, 2019.

[4] Oyama, Henrique, Dominic Messina, Renee O'Neill, Samantha Cherney, Minhazur Rahman, Keshav Kasturi Rangan, Govanni Gjonaj, and Helen Durand. 'Test Methods for Image-Based Information in Next-Generation Manufacturing.' *IFAC-PapersOnLine* 55, no. 7 (2022): 73-78.

268b Automated Outlier Detection and Estimation of Missing Data

Jinwook Rhyu, Dragana Bozinovski, Alexis B. Dubs, Naresh Mohan, Elizabeth M. Cummings Bende, Andrew J. Maloney, Miriam Nieves, Jose Sangerman, Amos E. Lu, Moo Sun Hong, Anastasia Artamonova, Rui Wen Ou, Paul W. Barone, James C. Leung, Jacqueline Wolfrum, Anthony Sinskey, Stacy Springs, Richard D. Braatz

Industrial process datasets commonly have missing values. The distribution of missing data points can be characterized as being within four classes: (1) random missing data, which exhibit no explicit pattern, (2) sensor drop-out, in which the missing values are correlated in time, (3) multi-rate, in which the missing data occur periodically, and (4) censoring, in which there exist thresholds for censoring so that the measurements outside the range are not recorded (Imtiaz and Shah, 2008; Severson et al., 2017).

The presence of missing values inhibits the use of the data in process modeling, analysis, and control. Even basic data analytics methods such as principal component analysis (PCA) and partial least squares (PLS) require a full data matrix, that is, without any missing values. The simplest way to deal with missing values is to only consider the observations with full measurements. Removing every observation with missing values, however, can cause significant data loss in the specific time period when the missing values are agglomerated, which makes capturing the process dynamics challenging. In order to fully utilize the given dataset, many general-purpose methods such as mean imputation, alternating algorithm, PCA data augmentation (Imtiaz and Shah, 2008), Bayesian PCA (Oba et al., 2003), singular value thresholding (Cai et al., 2010), and augmented Lagrange multiplier (Lin et al., 2010) have been developed to filling in missing values in a structured way, which are briefly reviewed in this presentation.

The aforementioned matrix recovery algorithms for filling in missing datapoints, however, are vulnerable to outliers. Not removing the outliers before filling in missing values emphasizes the effect of outliers and degrades the accuracy and reliability of results obtained by subsequent data analytics. These outliers can be detected by using the contribution map for T^2 and Q statistics (Miller et al., 1998; Chiang et al., 2000; Zhu and Braatz, 2014).

To the best of our knowledge, there has been no software that simultaneously detects outliers and fills in missing values in an automated way. In this presentation, we introduce an open-source software that automatically detects outliers, fills in missing values, and evaluates each algorithm used for matrix recovery. This presentation describes the framework of outlier detection and missing value estimation used in the software, and the demonstration of the software to data collected from a continuous biomanufacturing pilot facility at the Massachusetts Institute of Technology. The data are from the production of a monoclonal antibody produced by Chinese Hamster Ovary cells in a perfusion bioreactor.

To provide a thorough validation of the methods and software, the software is applied to a variety of datasets are varying types and extent of missing data constructed from an initial dataset in which all of the measurements are initially available. The performance of the various methods for filling in missing values are compared using several metrics including the normalized root-mean-squared error, the number of imputed values outside the boundaries, the number of imputed values considered outliers, and the computational time. Fifty simulations on each missing data scenario were conducted to obtain the distribution of these metrics. The matrix completion methods were the most effective except for the censoring case where probabilistic PCA methods were the most effective.

References

Cai, J. F., Candès, E. J., and Shen, Z. (2010). A singular value thresholding algorithm for matrix completion. *SIAM Journal on Optimization*, 20(4):1956–1982.

Chiang, L. H., Russell, E. L., and Braatz, R. D. (2000). *Fault Detection and Diagnosis in Industrial Systems*. Springer Science & Business Media, London, U.K.

Imtiaz, S. A. and Shah, S. L. (2008). Treatment of missing values in process data analysis. *Canadian Journal of Chemical Engineering*, 86(5):838–858.

Lin, Z., Chen, M., and Ma, Y. (2010). The augmented Lagrange multiplier method for exact recovery of corrupted low-rank matrices. arXiv:1009.5055.

Miller, P., Swanson, R. E., and Heckler, C. E. (1998). Contribution plots: A missing link in multivariate quality control. *Applied Mathematics & Computer Science*, 8(4):775–792.

Oba, S., Sato, M.-A., Takemasa, I., Monden, M., Matsubara, K.-I., and Ishii, S. (2003). A Bayesian missing value estimation method for gene expression profile data. *Bioinformatics*, 19(16):2088–2096.

Severson, K. A., Molaro, M. C., and Braatz, R. D. (2017). Principal component analysis of process datasets with missing values. *Processes*, 5(3):38.

Zhu, X. and Braatz, R. D. (2014). Two-dimensional contribution map for fault identification. *IEEE Control Systems*, 34(5):72–77.

268c Fusion of Eye-Tracking and Electroencephalography (EEG) Based Metrics for Estimating the Cognitive Workload of Control Room Operators in Process Industries

Mohd Umair Iqbal, Babji Srinivasan, Rajagopalan Srinivasan

Process industries are highly hazardous and accidents keep occurring in these industries. Thus, it is imperative to develop resilient safety barriers. While there are several layers of protection that prevent the hazards leading to accidents, human form on significant layer. This can be ascertained by the fact that majority of accidents in process industries occur because of human errors; human errors account for nearly 60-80% of process industry accidents (Brauer, 2016). A detailed analysis of severe industrial accidents that occurred in the period spanning 2008 to 2018 revealed that about 76.1% of these accidents can trace back their root cause to human failures (Jung et al., 2020). The advent of rapid digitization and concomitant automation has changed the nature of human-machine interaction, and hence the role of humans, in process industries. The nature of the operator's task has shifted from the one involving emphasis on perceptual-motor skills to the one involving cognitive activities such as monitoring, diagnosis, prognosis, decision making, and problem-solving (Pascual et al., 2019). These cognitive activities need to be carried out efficiently and correctly by an operator to achieve the design goals and deal with process abnormalities that are outside the outside the purview of automation systems. Inefficiencies in human cognitive performance often translate to high cognitive workload that can have catastrophic consequences in terms of compromising the safety in process industries. Cognitive workload is one of the important constructs of human performance and an increase in it leads to degradation in performance (Ghalenoei et al., 2022). This work addresses the critical need to estimate cognitive workload of process industry control room operators using a fusion of eye-tracking and electroencephalography (EEG) derived features.

Despite advances in human performance assessment, traditional methods often overlook critical cognitive aspects of operators' performance. Evaluation is typically based on metrics like success rates, deviations from standard operating procedures, and response times (Lee et al., 2000), which fail to fully capture the underlying reasons behind sub-optimal performance. Such underlying reasons may include cognitive workload, shortcomings in mental models and design of HMI. Advancement in sensor technology has made it possible for researchers to tap on physiological measurements for getting insights into the cognitive aspects of human performance, including cognitive workload. Physiological parameters reflect intrinsic mental state of an individual and provide critical insights into those aspects of cognitive behavior which are otherwise not directly observable such as level of attention, information acquisition pattern, thinking strategy, and workload. A variety of sensors have been developed to measure various physiological parameters (Zheng et al., 2014) and many of these are minimally invasive. These sensors primarily include eye tracking, and electroencephalography (EEG). Eye tracking provides critical insights into visual behavior. For instance, expert nuclear power plant operators have a different eye gaze pattern during stressful emergency accident situations, marked by higher total fixation durations compared to novices (Liu et al., 2019) thus providing critical insights into information acquisition pattern. Likewise, electroencephalography (EEG) provides critical information about operators' mental states, such as levels of cognitive workload (Liang et al., 2018). For instance, spectral power of EEG theta has been found to be sensitive to mental workload of air traffic controllers (Dasari et al., 2017). However, depending on a single sensor may not often be adequate in a real setting where factors such as measurement noise, and artifacts induced by a subject's movement can affect the accuracy and robustness of single sensor-based metrics. Moreover, some aspects of performance may not be sensitive to the metrics obtained from one sensor while the other sensor-based metrics can capture the same (Debie et al., 2021). Therefore, utilizing features from both sensors is beneficial. For instance, a classification accuracy of 93% for visual hazard detection was attained when features from eye-tracking and EEG were fused compared to accuracies of 73% (for eye-tracking) and 83% (for EEG) obtained independently (Noghabaei et al., 2021). Our eye-tracking research (Sharma et al., 2016; Bhavsar et al., 2017) and EEG studies (Iqbal et al., 2020, 2021) provide valuable insights into the cognitive behavior of control room operators. Experts have lower gaze entropy than novices, indicating more directed attention (Bhavsar et al., 2017). EEG can identify mismatches between an operator's mental models and actual process behavior (Iqbal et al., 2020).

In the current work, we study the added benefit of the fusion of eye-tracking and EEG for operator performance assessment. Ten participants were involved in the study, and each of them carried out several repetitive trials. During a typical trial, a participant has to carry out six tasks (each corresponding to a disturbance scenario). Overall, these participants performed 81 trials resulting in 438 task-participant pairs. During these tasks, participants' brain activity and eye gaze are recorded via a single electrode EEG sensor and Tobii 300 eye tracker, respectively. In addition, process data, alarm information, and operator actions are also recorded. We extracted EEG and eye tracking based metrics for each of the tasks to populate a feature matrix. These metrics include power spectral densities of pupil diameter in the range of 0-2 Hz, average pupil diameter, fixation duration-based (gaze) metrics and power spectral densities of EEG in the range of 1-30 Hz. We trained a decision tree-based model, based on several combinations of features, for estimating cognitive workload of operators. Our results indicate that fusion of metrics results in a marked improvement in assessing the cognitive workload of control room operators compared to using features from EEG or eye tracking alone. For instance, classification accuracy of the decision tree-based model increased from a minimum of 45% (using EEG based features) to a maximum of around 69% using a fusion of metrics obtained from EEG, pupil, and gaze data. The improvement in results is hypothesized to be due to differences in dynamics of EEG and eye data. We also developed an operator independent generic model for estimating cognitive workload during a task which achieved an accuracy of about 67%. The work has the potential to assess cognitive aspects of performance, and evaluate expertise during training, thus providing a comprehensive and robust performance assessment. This will help ensure safety and quality of product in process industries. In a broader perspective, it can guide the application of human factors for improving safety in industries by providing objective information about the cognitive aspects of performance.

Keywords: Process safety, Operator performance, Eye Tracking, Electroencephalography (EEG), Decision Trees

Acknowledgement: This work is partially funded by American Express Lab for Data Analytics,

Risk and Technology, Indian Institute of Technology Madras

References

Ghalenoei, M., Mortazavi, S. B., Mazloumi, A., & Pakpour, A. H. (2022). Impact of workload on cognitive performance of control room operators. Cognition, Technology & Work, 24(1), 195-207.

Bhavsar, P., Srinivasan, B., & Srinivasan, R. (2017). Quantifying situation awareness of control room operators using eye-gaze behavior. Computers & chemical engineering, 106, 191-201.

Brauer, R. L. (2016). Safety and health for engineers. John Wiley & Sons.

Diego-Mas, J. A., Garzon-Leal, D., Poveda-Bautista, R., & Alcaide-Marzal, J. (2019). User-interfaces layout optimization using eye-tracking, mouse movements and genetic algorithms. Applied ergonomics, 78, 197-209.

Iqbal, M. U., Shahab, M. A., Choudhary, M., Srinivasan, B., & Srinivasan, R. (2021). Electroencephalography (EEG) based cognitive measures for evaluating the effectiveness of operator training. Process Safety and Environmental Protection, 150, 51-67.

Iqbal, M. U., Srinivasan, B., & Srinivasan, R. (2020). Dynamic assessment of control room operator's cognitive workload using Electroencephalography (EEG). Computers & Chemical Engineering, 141, 106726.

Jung, S., Woo, J., & Kang, C. (2020). Analysis of severe industrial accidents caused by hazardous chemicals in South Korea from January 2008 to June 2018. Saf. Sci. 124, 104580

Lee, S., Jeong, I., & Il, M. (2000). Development of evaluation algorithms for operator training system. Computers & Chemical Engineering, 24(2-7), 1517-1522.

Liang, Y., Liang, W., Qu, J., & Yang, J. (2018, October). Experimental study on EEG with different cognitive load. In 2018 IEEE International Conference on Systems, Man, and Cybernetics (SMC) (pp. 4351-4356). IEEE.

Noghabaei, M., Han, K., & Albert, A. (2021). Feasibility Study to Identify Brain Activity and Eye-Tracking Features for Assessing Hazard Recognition Using Consumer-Grade Wearables in an Immersive Virtual Environment. Journal of Construction Engineering and Management, 147(9), 04021104.

Porthin, M., Liinasuo, M., & Kling, T. (2020). Effects of digitalization of nuclear power plant control rooms on human reliability analysis–A review. Reliability Engineering & System Safety, 194, 106415.

Shahab, M. A., Iqbal, M. U., Srinivasan, B., & Srinivasan, R. (2022). HMM-based models of control room operator's cognition during process abnormalities. 1. formalism and model identification. Journal of Loss Prevention in Process Industries, 104748.

Sharma, C., Bhavsar, P., Srinivasan, B., & Srinivasan, R. (2016). Eye gaze movement studies of control room operators: A novel approach to improve process safety. Computers & Chemical Engineering, 85, 43-57.

268d Next-Generation Manufacturing Considerations: Flow System Cybersecurity, Image-Based Control, Quantum Entanglement, and Learning Proofs

Keshav Kasturi Rangan, Henrique Oyama, Kip Nieman, Helen Durand

Next-generation manufacturing processes are faced with a range of considerations, ranging from cybersecurity to image-based sensing to automating reasoning. Examples of directions in next-generation manufacturing systems have included designing control systems which attempt to detect cyberattacks [1], utilizing quantum computation as part of scheduling [2], or incorporating novelty detection with convolutional neural networks for image-based control [3]. Motivated by the broad range of topics of relevance to next-generation manufacturing, in this talk, we overview several of our recent results throughout the broad range of next-generation manufacturing-relevant topics.

We first review our recent results related to cybersecurity of control systems for flow in a pipe [4] . Using ANSYS Fluent as a simulator for the flow in the pipe and modifying the flow with a user-defined function, we look at how different sensor placements along the pipe length can impact how quickly an abnormal event (which could be representative of a cyberattack) is noted. Specifically, if a temperature sensor is placed in the middle of the pipe system, we can expect that the time at which the temperature changes can be observed would be halved, as the fluid would take half the time to travel half the pipe. From this view, which involves dealing with process states that vary with respect to time and space, we see that the closer the temperature is measured to the source of the attack, the faster one can signal the abnormality. Though placing more sensors in different locations of the system may have the benefit of fast attack detection, it is important to ask what the minimum security architecture for sensor placement is to detect attacks as fast as possible without significantly adding more costs. In the next part of the talk, we will discuss image-based control, and will review how simulations in Blender can be used to represent image-based controllers [5]. Next, we will discuss why the limitations of quantum entanglement (and specifically that entanglement alone does not permit signaling faster than the speed of light) cannot be overcome through strategies such as state predictions to facilitate useful communication in next-generation manufacturing [6]. Finally, we analyze what the outputs of a language model trained using proofs of Lyapunov-based economic model predictive control (LEMPC) that have been tailed for applications in areas such as cybersecurity of control systems (e.g., [7]). We will discuss what the results are, and why they are the results, to discuss the potential of this method to be used in guiding proofs for process systems engineering-relevant topics.

References:

[1] Narasimhan, S., El-Farra, N. H., & Ellis, M. J. (2022). A control-switching approach for cyberattack detection in process systems with minimal false alarms. *AIChE Journal*, *68*(12), e17875.

[2] Ajagekar, A., Al Hamoud, K., & You, F. (2022). Hybrid classical-quantum optimization techniques for solving mixed-integer programming problems in production scheduling. *IEEE Transactions on Quantum Engineering*, *3*, 1-16.

[3] Pulsipher, J. L., Coutinho, L. D., Soderstrom, T. A., & Zavala, V. M. (2022). SAFE-OCC: A novelty detection framework for Convolutional Neural Network sensors and its application in process control. *Journal of Process Control*, *117*, 78-97.

[4] Kasturi Rangan, K., Oyama, H., Azali Assoumani, I., Durand, H., & Ng, K. Y. S. (2023).

Cyberphysical Systems and Energy: A Discussion with Reference to an Enhanced Geothermal Process. In Energy Systems and Processes: Recent Advances in Design and Control (pp. 8-1). Melville, New York: AIP Publishing LLC.

[5] Oyama, Henrique, Dominic Messina, Keshav Kasturi Rangan, Akkarakaran Francis Leonard, Kip Nieman, Helen Durand, Katie Tyrrell, Katrina Hinzman, and Michael Williamson. 'Development of directed randomization for discussing a minimal security architecture.' *Digital Chemical Engineering* 6 (2023): 100065.

[6] Nieman, K., Kasturi Rangan, K., & Durand, H. (2022). Control Implemented on Quantum Computers: Effects of Noise, Nondeterminism, and Entanglement. *Industrial & Engineering Chemistry Research*, *61*(28), 10133-10155.

[7] Oyama, Henrique, Dominic Messina, Keshav Kasturi Rangan, and Helen Durand. 'Lyapunov-based economic model predictive control for detecting and handling actuator and simultaneous sensor/actuator cyberattacks on process control systems.' *Frontiers in Chemical Engineering* 4 (2022): 810129.

268e Single-Step Synthesis of Shaped Polymeric Particles Via Initiated Chemical Vapor Deposition By Leveraging Liquid Crystals As a Reaction Medium and Real-Time Display

Apoorva Jain, Soumyamouli Pal, Nicholas L. Abbott, Rong Yang

Shaped polymer particles will benefit a wide range of applications such as targeted drug delivery vehicles and in soft robotics with reconfigurable super assemblies and directed motion, but existing synthesis approaches are commonly multistep and limited to a narrow size/shape range. In this study, liquid crystals (LCs) were leveraged both as a reaction medium and as a real-time display of the polymerization conditions and progression, by setting up a custom in-situ long-focal range microscope assembly. Tapping into the anisotropy in molecular alignments of nematic liquid crystals (LCs), we developed a novel single-step technique for polymerizing divinylbenzene (DVB) via initiated chemical vapor deposition (iCVD) to generate a variety of shapes including nanospheres, hemispherical micro-domes, orientation-controlled microgels, microspheres, spheroids, and micro-discs. In-situ monitoring coupled with iCVD allowed for precisely optimized and continuous delivery of vapor-phase reactants, thus avoiding disruption of the LC structure, a critical limitation in past LC-templated polymerization. In addition, by implementing cross polarized microscopy characterization in real-time, we unraveled the role LC-substrate interface played in the formation of shaped particles. Detailed image analysis and chemical composition characterization revealed key mechanistic insights of shape and size controlled polymer synthesis in LC. By harnessing LC display output as signal amplifiers for monitoring reaction microenvironment in real-time in this study, we lay the foundation for agile development of next generation particle synthesis technologies, to be further employed in tandem with data-driven approaches.

268f Control Strategies Using Quantum Computations for Next-Generation Manufacturing

Keshav Kasturi Rangan, Helen Durand

Since Industry 4.0 standards were introduced, the importance of collecting and processing data has skyrocketed to ensure that industries not only improve their process efficiency, but also transparency. Chemical processing demands are increasing along with those in multiple other fields. Quantum computation, as a new computing technique, is, as a result, receiving interest in a variety of fields [1], including process systems engineering [2]. From a controls perspective, however, the utility of quantum computing remains unclear. However, we hypothesize that by implementing controllers on quantum computers (initially starting with simpler control laws and then progressing to more advanced control laws), we will be able to gain insights into how and whether quantum computation could be useful for process control.

With this motivation, the concept of quantum computing from a controls perspective is investigated by controlling a simple linear system using a proportional (P) controller implemented in a hybrid quantum-classical fashion [3], where classical pre-processing is used to prepare data for summation to create the result of the multiplication. This is achieved by using a Quantum Fourier Transform-based [4] addition algorithm on IBM's quantum experience SDK, Qiskit. This leverages the code in [7] for addition with the quantum computer. The algorithm will then be compared to an implementation of the same control strategy as a quantum simulation that does not use the same classical pre-processing to analyze the impact of quantum noise on different algorithm structures. A variety of factors that influence the results will be analyzed, including the process states measured, number of shots and quantum register size used to represent process states [3,5,6]. We will discuss noise models and how the noise models affect the theory and practice of quantum computing-implemented control.

References:

[1] Deng, Z., Wang, X., & Dong, B. (2023). Quantum computing for future real-time building HVAC controls. Applied Energy, 334, 120621.

[2] Ajagekar, A., & You, F. (2020). Quantum computing assisted deep learning for fault detection and diagnosis in industrial process systems. *Computers & Chemical Engineering*, *143*, 107119.

[3] Keshav Kasturi Rangan et al. "Quantum Computing and Resilient Design Perspectives for Cybersecurity of Feedback Systems". In: IFAC-PapersOnLine 55.7 (2022), pp. 703–708

[4] Ruiz-Perez, L., & Garcia-Escartin, J. C. (2017). Quantum arithmetic with the quantum Fourier transform. Quantum Information Processing, 16(6), 1-14.

[5] Nieman, K., Kasturi Rangan, K., & Durand, H. (2022). Control Implemented on Quantum Computers: Effects of Noise, Nondeterminism, and Entanglement. Industrial & Engineering Chemistry Research, 61(28), 10133-10155.

[6] Kasturi Rangan, K., Oyama, H., Azali Assoumani, I., Durand, H., & Ng, K. Y. S. (2023). Cyberphysical Systems and Energy: A Discussion with Reference to an Enhanced Geothermal Process. In Energy Systems and Processes: Recent Advances in Design and Control (pp. 8-1). Melville, New York: AIP Publishing LLC.

[7]Anagolum, S. DoNew. 2018. https://github.com/SashwatAnagolum/DoNew.

268g Smart Product and Process Design – a Digital Thread Approach to Enable Consumer Packaged Goods (CPG) Companies' Transformation to a Smart and Adaptable Enterprise

Olivia Musto, Ravindra Aglave, Nadia Naraine, Steven Mallia, Veronica Tchernychouk, Erin Mitchell

Consumer Packaged Goods (CPG) companies, specifically food & beverage and cosmetics companies, today face constantly changing and increasingly complex market trends driven by consumer habits. As a result, companies have replaced siloed systems with digital platforms that allow them to be responsive and build a 2-way relationship with consumers.[1] Despite that, 95% of new product launches per year fail.[2] Therefore, companies face the challenge of developing connected, flexible, adaptable, and sustainable products that will drive customer loyalty and improve their bottom line. The solution is digital transformation using a "digital thread" approach for smart product and process design; an approach to design and virtual validation leveraging simulation and concurrent design. Such approach ensures manufacturability at desired speed, quality, and cost.[3]

A "digital thread" is a discrete, linked, traceable sequence of activities in the product or production lifecycle, that is digitalized and automated.[4] For the food and beverage industry, Siemens outlines 5 digital threads, shown below, which comprise the entire digital twin and are implemented as a complete solution set with the Xcelerator portfolio.

Figure 1

Each thread is further characterized to detail the components that implement the solution. The smart product and process design (SPPD) thread is one such thread, the steps from which are shown below.

Figure 2

1. Beginning with formula design, integrated with laboratory testing, the formulation technology and computer algorithms work together to uncover optimal quantities of raw materials and minimize cost while maintaining nutritional and quality requirements.[5]
2. Laboratory quality control includes automated contextualization of test batch information and results validation to establish quality standard, protect the product design and enforce regulatory requirements.[5]
3. Package design is done in a centralized collaboration platform to maximize re-use and utilizes detailed simulations to ensure package integrity during packaging, e-commerce transport and consumer handling.
4. Artwork and labeling is seamlessly integrated in the collaboration platform to allow immediate data exchange across disciplines and ensure data accuracy for every label.[5]
5. Recipe planning leverages a digital set of capabilities that harness the complexity of recipes by automatically modifying them for scale-up across plants based on pre-defined rules.[5]
6. Recipe simulation incorporates Multiphysics simulation of plant-specific equipment to virtually validate manufacturability and optimize the recipe.[5]
7. Packaging process planning bridges the gap between R&D and manufacturing by using a standardized method to define filling and packaging processes and creating straightforward EWIs to ensure rapid scale-up.[5]
8. Finally, packaging process simulation validates the manufacturability and optimizes filling and packaging processes using Multiphysics simulation with the plant site's specific equipment and layout.[5]

With this multi-disciplinary product design, a digital twin of both the product and production is generated to provide a complete set of solutions for a product maintaining its connection to the manufacturing process. This results in an integrated environment where all change drivers can be considered simultaneously. Process design and planning are performed to implement product development capabilities, maximizing reuse and increasing collaboration by connecting processes with a single source of data.[3]

Recipes can be scaled-up faster and optimal operating setpoints are uncovered; currently a speed limiting step. Furthermore, the manufacturing process validation verifies that products can be manufactured, assembled, and packaged virtually to reduce the time and cost of physical testing. Creating a site-specific manufacturing plan enables the optimization of the filling and packaging processes. Utilizing the "Smart Product and Process Design (SPPD)" digital thread approach, companies can bring new products to market ~15-35% faster.[3] This allows them to meet consumers' demand for personalized experiences and enter the market at the right time while keeping the costs low.

In this presentation, we will present the steps involved in the "Smart Product and Process Design" digital thread with the help of examples and recorded demos for the use of case of cold brew coffee. It will help establish an understanding of the procedures and widely growing benefits of digital transformation.

[1] Marfa, E., "Smart Product and Process Design Executive Presentation Recording," *Highspot* Available: https://siemens.highspot.com/items/63fe51caaa621aa2caa5cdd3?lfrm=srp.5.

[2] "Product innovation: 95% of new products miss the mark," *MIT Professional Education* Available: https://professionalprograms.mit.edu/blog/design/why-95-of-new-products-miss-the-mark-and-how-yours-can-avoid-the-same-fate/#:~:text=Nearly%2030%2C000%20new%20products%20are,Coca%2DCola%2C%20and%20Colgate.

[3] Marfa, E., "Smart Product & Process Design - Food & Beverage," *Highspot* Available: https://siemens.highspot.com/items/60cc6b683f65f6508db05dc2?lfrm=shp.0.

[4] Jackson, C., "Digital Threads and business processes - siemens xcelerator software for industry," *Xcelerator* Available: https://blogs.sw.siemens.com/xcelerator/2021/08/20/digital-threads-and-business-processes/.

[5] " Digital Enterprise Touch App - Focus F&B with Digital Threads HM21," *Highspot* Available: https://siemens.highspot.com/items/62bc5c47070e798f4e41f4fd?lfrm=shp.0#34.

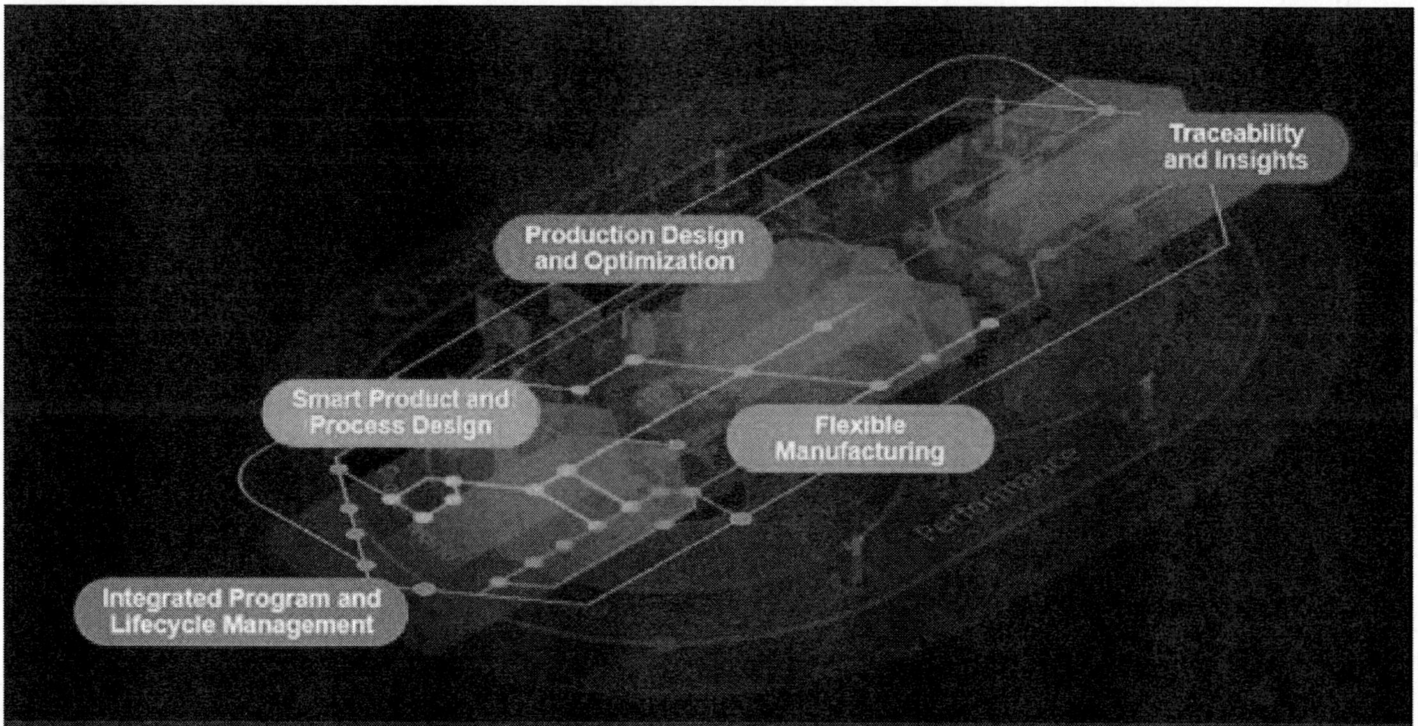

Fig. 1 Food & beverage digital threads.

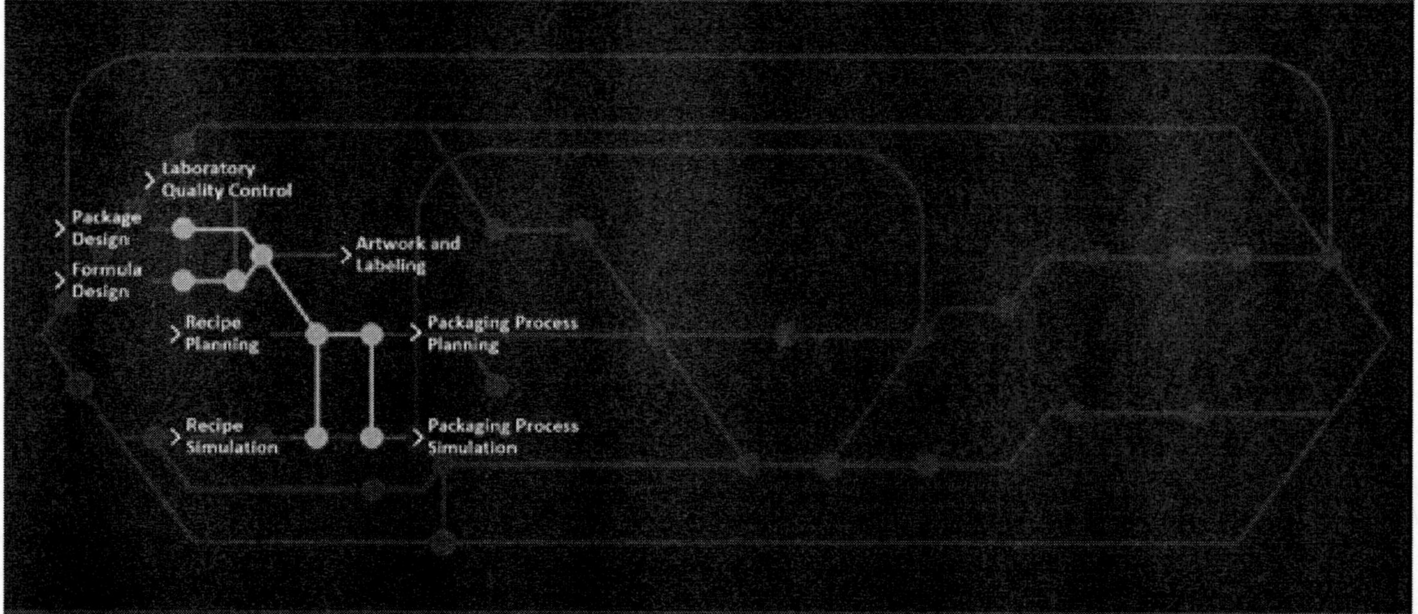

Fig. 2 Smart Product & Process Design digital thread steps.

69a Designing Efficient Human-Machine Interfaces for Decision-Support Tools: Case Studies in Healthcare and Process Systems

Mrunal Sontakke, Sambit Ghosh, Faye Cameron, Alan Ganz, Henry Weber, Lucky E. Yerimah, Andreas Rebmann, Craig Dory, Ronald Hedden, Joel Plawsky, Johnson Samuel, B. Wayne Bequette

Human-machine interfaces link human decision-making with the underlying processes, measurements from the process, and the algorithms used for data analysis. They enable humans to interact with complex systems and make sense of the information they provide, allowing us to make informed decisions. The type of data feedback dramatically influences our decision-making. Incorrect decisions increase risk exponentially, specifically for safety-critical domains like healthcare and process systems. This article presents the design of efficient human-machine interfaces required at different stages of developing decision-support tools in these two domains.

Healthcare: We focus on Type-1 diabetes, a condition where the pancreas ceases to produce insulin requiring external insulin input to control blood glucose. Developing decision-support systems can ease the cumbersome task of continuous monitoring by patients. However, comprehensive context-driven data is needed to develop the algorithms required for decision-support systems. Required data can be collected from patients or caregivers in free-living conditions. To make this task effortless, we designed a human-friendly interface for daily activity (eating, sleeping, exercising), logging on to smartphones, and collecting accelerometer data from smartwatches.

Process systems: This case study focuses on small and medium-scale food industries lacking digitalization and real-time quality monitoring. To address this issue, new sensors were installed on the output line to monitor and collect real-time product stream data, improving efficiency. The human-machine interfaces were designed to provide clear and easily understandable information about the sensor output and decision support if necessary. The data transfer was enabled using cloud platforms, specifically Smart Manufacturing and Innovation Platform (SMIP) developed by the Clean Energy Smart Manufacturing Innovation Institute (CESMII).

In human-in-the-loop decision-making, one must understand how humans respond to safety-critical operations like start-ups and shutdowns. A plantwide dynamic process simulation of vinyl acetate monomer production is developed, with the human-machine interface providing the necessary inputs to the operator. The simulation allows for safely testing various scenarios and provides an opportunity to study the operator's responses and understand human-in-the-loop decision-making. The human-machine interface is designed so the operator can get decision support during critical start-up and shutdown operations. By tracking operator responses, guidelines for efficient interface design can be developed.

The gap between industry and academia has long been reported and discussed in the literature. As a step toward reducing this divide, a pedagogical human-machine interface and guidance system are developed for a heat exchanger used by undergraduate students. The system uses cloud platforms for sharing data. The interface is designed to help students understand the experiment process dynamics by providing multi-modal information about the experiment in real time. This real-time understanding of the process dynamics helps students make correct manual decisions for successful experiment learning. Furthermore, the interface enables interaction and synthesis with student-designed fault detection algorithms. This provides a realistic setting to deploy algorithms for real-time decision support and better prepares them for careers in industry, where they will encounter similar challenges with real-world equipment and data.

Briefly, the proposed presentation aims to educate viewers on machine interfaces in human-in-the-loop systems, including the importance of considering the needs of individuals based on their level of expertise and modifying designs accordingly.

69b GMP Implementation of Continuous Manufacturing: A Case Study

Edward Wong

Continuous processing is being explored in various ways throughout the biopharmaceutical industry. New process technologies and business considerations are driving process design towards more agile, modular, and highly productive paradigms. Continuous manufacturing provides potential benefits such as reduced process scale and the implementation of single-use technologies. However, these benefits may come at the cost of complexity.

This presentation will review a case study analyzing the conversion of a traditional GMP process into a continuous manufacturing paradigm. Discussion will include the development of integrated continuous operations, control strategy, and process design considerations. The outcomes of the process conversion will also be discussed.

59v Machine Learning-Based Prediction and Optimization of Liquid Wettability of an iCVD-Produced Fluoropolymer

Daniel Schwartz, Tien Nguyen, Zhengtao Chen, Kenneth Lau, Michael C. Grady, Ali Shokoufandeh, Masoud Soroush

Abstract

Initiated chemical vapor deposition (iCVD) is a reactive process that creates polymeric materials on a surface from vapor-phase monomers and thermal initiators. Our iCVD synthesis of poly(perfluorodecyl acrylate) (PPFDA) resulted in the growth of micro- and nano-worms normal to the surface [2, 3]. The micro- and nanostructures of the worms directly depend on iCVD process conditions. They in turn influence bulk properties such as their liquid wettability. The current absence of a physicochemical model that can explain the relationships between iCVD process conditions and bulk properties of the polymers (e.g., contact angles of liquids on the polymer surface) motivates the use of data-driven modeling to capture and describe the relationships.

In this talk, we: (i) report iCVD data (contact angles of heptane, octane, and water on PPFDA and process conditions) from 49 batches; (ii) capture and describe the relationships using artificial neural networks; and (iii) use the models to determine the optimal iCVD process conditions that maximize the contact angles on PPFDA [4]. We experimented with different machine learning models to score high accuracy with the limited training data available for modeling the relationships between iCVD conditions and contact angles. Research has shown that, in general, training a classifier from scratch on small datasets does not work well. Recently, Barz *et al.* showed that using a cosine loss function provides better performance than traditional objective functions on smaller datasets [6]. We studied the validity of the hypothesis of using cosine loss on a multi-layer perceptron trained on few data samples. We also studied a second hypothesis that convolutional layers lead to higher order features that improve the predictions. We implemented a convexified convolutional neural network that approximates the best possible convolutional neural network and can estimate the optimal conditions in a more efficient and explanatory method using convex optimization. The ability to create surfaces with high contact angles and liquid repellency is important for a broad range of applications, such as in oil-water separations, self-cleaning surfaces, and microfluidics [5].

Keywords

Machine learning, data-driven, iCVD polymers, liquid repellency, processing-condition-property relationship

References

[1] F. Civan, "instrumental and laboratory techniques for characterization OF reservoir rock," in *Reservoir Formation Damage*: Gulf Professional Publishing Burlington, 2007, pp. 154-173.

[2] K. K. S. C. Lau, Z.; Nguyen, T. , "Directed Vapor Deposition and Assembly of Polymer Micro- and Nanostructures. U.S. Patent Application 62941086," November 27, 2019 2019.

[3] Z. N. Chen, T.; Lau, K.K.S. , "One-Step Bottom-Up Growth of Highly Liquid Repellent Worm-Like Surfaces on Planar Substrates, ACS Nano (submitted)." 2021.

[4] D. Schwartz *et al.*, "Data-driven prediction and optimization of liquid wettability of an initiated chemical vapor deposition-produced fluoropolymer," *AIChE Journal*, vol. 68, no. 6, p. e17674, 2022.

[5] A. K. Kota, G. Kwon, and A. Tuteja, "The design and applications of superomniphobic surfaces," *NPG Asia Materials*, vol. 6, no. 7, pp. e109-e109, 2014.

[6] Barz B, Denzler J. Deep learning on small datasets without pre-training using cosine loss. Paper presented at: Proceedings of the IEEE/CVF Winter Conference on Applications of Computer Vision2020.

69d Towards Scalable and Cost-Effective Plasmid DNA Manufacturing

Niki Triantafyllou, Miriam Sarkis, Nilay Shah, Maria Papathanasiou, Cleo Kontoravdi

Advanced Therapy Medicinal Products (ATMPs) are a novel class of therapeutics that have demonstrated promising results in the cure of life-threatening diseases, such as cancer and neurodegenerative disorders. ATMPs have attracted global interest that translates into 27 approved products, 1,454 ATMP manufacturers worldwide, and 2,220 active clinical trials, as of 2022 [1, 2]. The adoption of scalable and cost-effective manufacturing processes is key for meeting the unprecedented demand for these products. Plasmid DNA (pDNA) is a critical raw material used as carrier of the gene of interest [3]. Currently, pDNA supply is limited and costly and the demand is expected to increase rapidly over the coming years, mainly due to its key role in the manufacturing of cell and gene therapies and mRNA and viral vector vaccines [4]. Scaling up the pDNA manufacturing process to meet current and future demand is one of the main challenges of the ATMP sector [4].

In this work, we employ Process Systems Engineering (PSE) methodologies as an approach to assist systematic and proactive decision-making throughout the stages of process design and optimization. We propose an integrated computational framework to aid wider adoption of ATMPs by ensuring efficient timelines, cost-effectiveness, and scalability in the production of pDNA. Specifically, we identify key process and economic parameters for the production of pDNA and develop techno-economic models for different scales, phase I clinical trials to commercialization [6]. Through uncertainty and global sensitivity analysis, we quantify key process and economic uncertainties in the manufacturing process (Figure 1a) and apportion them to key process parameters (Figure 1b) [5]. We thus highlight cost drivers and limitations that steer decision-making, as well as the importance of economies of scale in the production of pDNA. Finally, we perform Bayesian optimization to identify optimal sets of process parameters that lead to reduced costs, in an attempt to decrease the cost of goods (COGs), improved lead times, increased productivity, and reduced the environmental footprints of the process [7]. Optimized manufacturing recipes identified with the proposed framework can achieve up to a 170% increase in batch size and a 34.7% decrease in the operating cost per batch (Figure 1a).

Figure 1. (a) Batch size variability for the production of pDNA in a commercial scale. The black bar in the center is the interquartile range, the white dot inside the black bar is the median value, and the black lines stretched from the bar are the lower/upper adjacent values. Bayesian optimization is used to maximize the batch size. (b) Global sensitivity analysis of the nine critical input parameters on the batch size. First-order effects () are displayed in the diagonal and second-order effects () are presented in the upper and lower triangular.

References

[1] Hunt TD (2023) CELL & GENE: state of the industry briefing. https://alliancerm.org/arm-event/sotibriefing/. Accessed 28 Feb 2023

[2] The TransMission: Current FDA Approved Gene and Cell Therapies. https://www.mirusbio.com/blog/fda-approved-gene-cell-therapies. Accessed 27 Feb 2023

[3] M. Gosse, C. Jones, D. Jesus, S. D'Costa, 2022, Regulatory & supply chain implications for plasmids as critical starting materials in the manufacture of viral vector gene therapy products. Cell and Gene Therapy Insights, 8(2), 279–286.

[4] J. Ohlson, 2020, Plasmid manufacture is the bottleneck of the genetic medicine revolution, Drug Discovery Today, 25(11), 1891–1893.

[5] S. Kucherenko, 2013, SobolHDMR: A general-purpose modeling software. Methods in Molecular Biology, 1073, 191–224.

[6] R. Ferreira, D. Petrides, 2021, Plasmid DNA (pDNA) Large Scale Manufacturing – Process Modeling and Techno-Economic Assessment (TEA) using SuperPro Designer, doi: 10.13140/RG.2.2.12780.28800.

[7] M. Balandat, B. Karrer, D. R. Jiang, S. Daulton, B. Letham, A. Gordon Wilson, E. Bakshy, 2020, BoTorch: A Framework for Efficient Monte-Carlo Bayesian Optimization, Advances in Neural Information Processing Systems, 33.

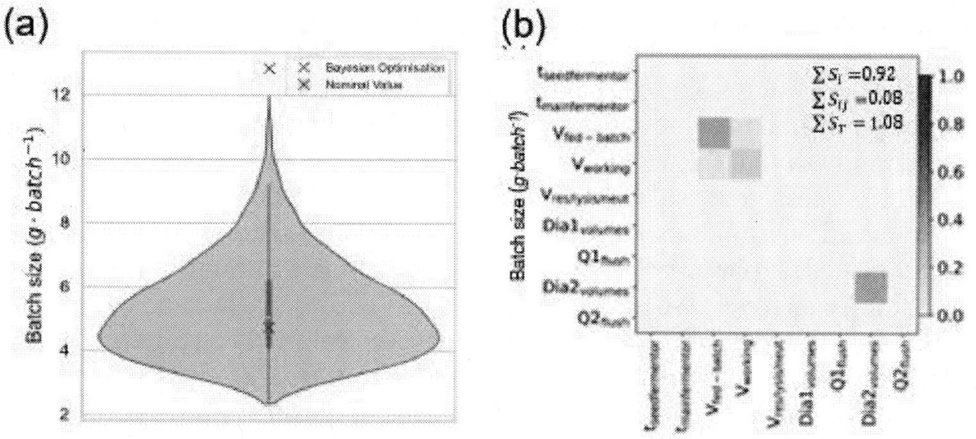

69e Barrier-Free Paper Analytical Devices for Multiplex Colorimetric Detection

Ayushi Chauhan, Bhushan J. Toley

Microfluidic paper-based analytical devices (μPADs) are being increasingly adopted to perform colorimetric assays for various applications. The reason is attributed to their portability, biocompatibility, and pump-free designs. Multiplex colorimetric detection in μPADs is achieved by patterning hydrophobic barriers (Martinez et al., 2007). These barriers were created using wax printing, inkjet printing, photolithography, or chemical modification of paper. Although these fabrication methods efficiently utilize membrane surface area and consume less reagent volume, they require expensive equipment and skilled personnel, posing a challenge for scale-up. In addition, the convection of rehydrated reagents towards membrane edges produces non-uniform signals, which make result interpretation difficult. To overcome the limitations of conventional μPADs, we present a new device called barrier-free μPAD (BF-μPAD) that can detect multiple targets without patterning barriers, enabling faster and homogenous signal generation. In one embodiment, an 8cm x 2cm device assembly can perform up to 20 different tests in 30 seconds. A user only requires a smartphone to visualize and interpret signals, making it a desirable point of care tool at low resource settings.

BF-μPADs consists of a stack of two paper layers of different wicking rates. The top layer is composed of Standard 17 glass fiber (called distribution layer) and offers less resistance to fluid flow. The bottom layer is composed of Whatman filter paper Grade 1 (called detection layer) and offers high resistance to fluid flow. The assembly was secured using acrylic supporting layers. Test reagents were deposited on the detection layer, and samples containing different analytes were added through the distribution layer. After the signals were fully developed, the device was flipped, and the colored signals were visualized using a smartphone in a custom-made imaging box and using a flatbed scanner. The image was then split into RGB color channels, and intensities of each spot were measured in the suitable channel using ImageJ software.

Prior to multiplex detection, a preliminary study was performed to investigate the uniformity of signals on the detection layer. 16 spots of bromophenol blue, a pH indicator dye, were deposited on an 8cmX2cm detection layer (Fig. 1A). An equal dimension of Standard 17 was placed flush on the detection layer, and 895ul (volumetric capacity of distribution and detection layer) of 1M HCl was added as the rehydrating fluid through the distribution layer. On addition of the acid, the pH of the dye changes, and the blue spots turn yellow. Figure 1B shows the colorimetric signals developed in the detection layer. It was observed that the intensities of the yellow color were similar and independent of the location of spots from the fluid source (Fig. 1C). The coefficient of variations of 48 spots across three device replicates was 6.94% for the smartphone and 3.49% for the scanner.

The multiplexing feature of BF-µPAD was demonstrated by testing salivary thiocyanate, glucose, nitrite, and bovine serum albumin (BSA) in a 2cm x 2cm assembly. Figure 2A shows the user experience of the device. After saturation of signals, the device was flipped, and the image of the detection layer was acquired using a smartphone. Figure 2B shows the colorimetric signals obtained in each spot for a sample containing 1 mM SCN-, 10 mg/dl glucose, 0.8 mM NO2- and 3 mg/ml BSA. Eight sample fluids with different concentrations of the four analytes were added to BF-µPAD, and mean test spot intensities (N=3 devices) were plotted against analyte concentrations. Intensities of all analytes increased monotonously with concentrations. Figure 2C shows the calibration plots of the four analytes.

We also compared the limit of detection (LOD) of the BSA assay performed in BF-µPAD and a traditional single-layer µPAD using a statistical method described by Holstein et al (Holstein et al., 2015). LOD and the corresponding 95% confidence interval obtained using BF-µPAD were 0.1817 mg/ml and 0.1283-0.3682 mg/ml, respectively. The corresponding figures for µPAD were 0.6727 mg/ml and 0.5314-0.8217 mg/ml, respectively, signifying that BF-µPAD improved the LOD of the colorimetric assays by 3.7x compared to conventional µPAD. Further, a comprehensive study was performed to investigate the cross mixing of rehydrated signals on a longer assembly by depositing additional reagents upstream of the test signals. It was observed that the rehydrated signals from one chemistry do not mix with the downstream signals.

Barrier-free colorimetric detection in BF-μPAD is achieved by the unique flow pattern. The distribution layer rapidly wicks fluid in the lateral direction and continuously supplies it to the bottom detection layer from the vertical direction. Vertical entry of fluid practically eliminates lateral fluid movement in the detection layer, obviating the need for patterning barriers and generating spatially uniform signals. These flow patterns were modeled in COMSOL Multiphysics using Richards equation but are not presented here for brevity. Although the application of BF-μPAD was shown for biological fluid, the device can also be adapted to detect different pollutants in the environment, contaminants in soil, and pesticides in food. Since the distribution layer has large pores, the assembly is well suited for testing viscous emulsions like milk and denser fluids like honey which are challenging to flow alone through the cellulosic membrane of conventional μPADs.

BF-μPADs avert the need to pattern paper and enable faster flow and generate spatially uniform signals, improving upon the LOD of the assay. In contrast to μPADs, large-scale manufacturing of BF-μPADs would only require a commercially available benchtop robotic dispensing system to handle microliter volume.

69f Advanced Manufacturing Via Self-Assembled Colloidal Cracking of Binary Nanoparticles

Ryan Dumont, Bo Li

To face formidable challenges of future electronics with complex functionality and architecture, not only new materials but also new fabrication method need to be developed. However, topographical micropatterns (such as microchannels) are conventionally fabricated by complicated lithographic method, which are usually expensive and time-consuming, and not suitable for high throughput manufacture. Therefore, a low-cost, large-scale method for topographical template preparation needs to be developed. Recently, cracks formed in drying colloidal film have been demonstrated as ideal materials for microscale manufacturing. In this work, we first studied the key parameters that affect the dimensions of colloidal microchannel, such as channel-to-channel distance, channel width, and theoretical model is proposed to provide insights on colloidal microchannel formation. Then, the colloidal microchannel formation of binary particle solution was explored. Importantly, the colloidal microchannel formation mechanism will be studied. This work will provide a general strategy for generating of large scale microchannels, thereby offering a low-cost and lithography-free method for fabrication of patterns for printed electronics.

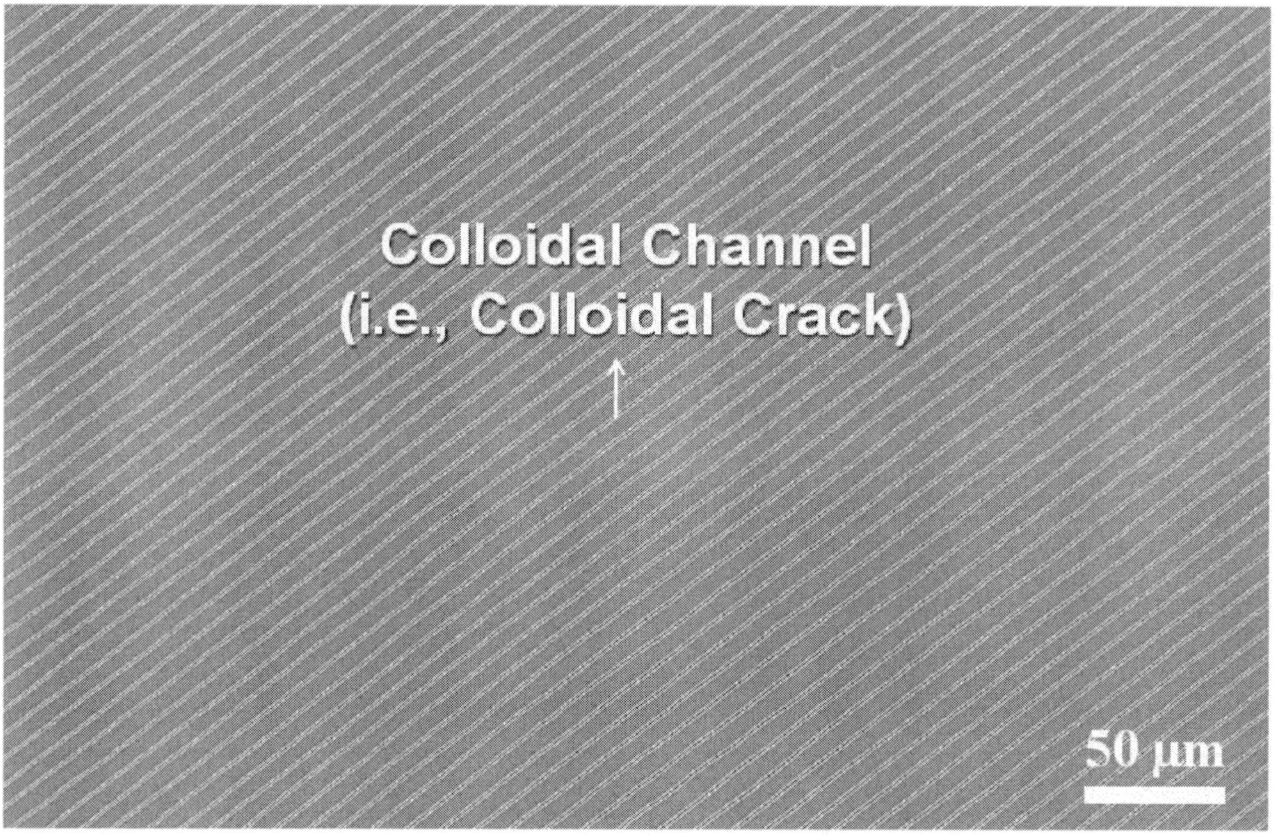

69g Bioprocess Control Using Stoichiometric Models of Metabolism

Mariana Monteiro, Sarah Fadda, Cleo Kontoravdi

Mammalian cells produce up to 80% of the commercially available therapeutic proteins, including monoclonal antibodies, with Chinese Hamster Ovary (CHO) cells being the primary production host. Manufacturing involves a train of reactors, the last of which is typically operated in fed-batch mode, where cells grow and produce the required protein. The reactor's product yield is typically low (up to 200 mL L^{-1} day^{-1}), which complicates downstream purification, accounting for up to 80% of the total production cost. Commonly employed optimization strategies include optimizing specific cell culture conditions (temperature and pH) and the feeding strategy. The feeding strategy is often decided *a priori*, from either prior knowledge or through design of experiments, and rarely considers the current state of the process. Advanced process control strategies like model predictive control are not typically used as online monitoring of samples is limited. In this work, we propose an advanced control strategy that leverages a metabolic model to optimize process yield and that requires only offline measurements commonly available for most fed-batch processes.

The model predictive control formulation proposed herein is built on a generic hybrid kinetic-stoichiometric reactor model and is therefore agnostic to particular cell lines or products. In this sense, the model is readily transferable across CHO cell culture systems. The control strategy (graphically represented in the image below) leverages a reduced metabolic network model of a CHO cell to calculate glucose and amino acid uptake rates that will lead to optimal cell growth for each culture phase. These fluxes are then used as control targets to compute the optimal feeding strategy. The optimal feeding strategy is then simulated in a high-fidelity model of the production reactor to predict the outcome *in silico*. We compare our closed-loop results against experimental data of a feeding strategy designed for antibody maximization [1]. Our results suggest that antibody production can be increased even further.

The proposed strategy is a first step towards integrating Industry 4.0 tools in bioprocessing. It does not require extensive computational power and can be implemented without extra infrastructure costs. It also bypasses the need to develop a complete dynamic metabolic model for CHO cells, which would be computationally challenging.

References:

[1] Kyriakopoulos S, Kontoravdi C. A framework for the systematic design of fed-batch strategies in mammalian cell culture. *Biotechnol Bioeng*. 2014;111(12):2466-2476. doi:10.1002/bit.25319

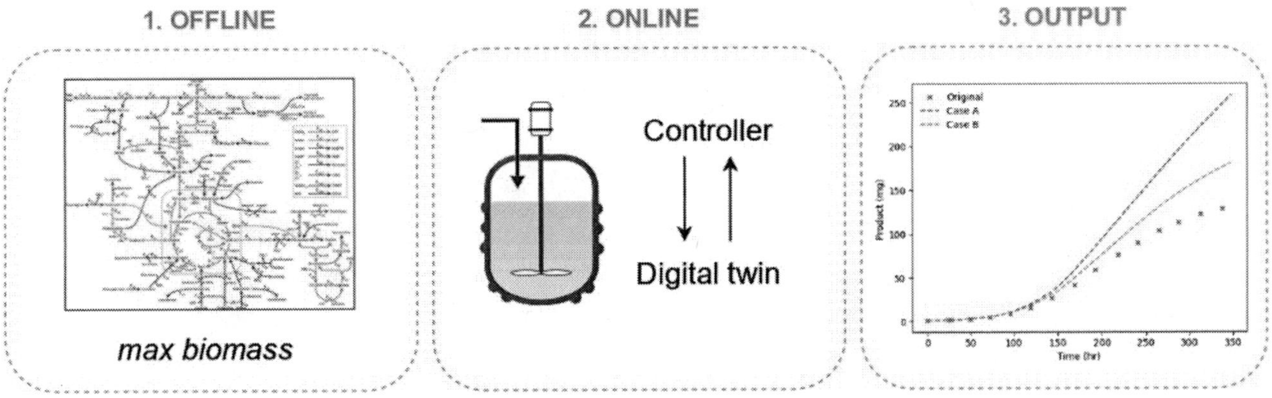

69h Edge-Enabled Monitoring System in 3D Printing Manufacturing Factory

Michelle C. Almendrala, Danielle Jaye Agron

Recent studies present smart monitoring programs to increase the mobility of the 3D printing technology. These monitoring systems is costly to implement in a multiple printers in a factory-level facility. To cope with it, a smart monitoring system enabled by edge devices were developed using LabVIEW program for multiple FDM printers operating inside the 3D printing factory. Edge devices were utilized to collect the data from the group of sensors connected to each individual 3D printer and to control the light modules that indicate the fabrication status. From there, these data were transmitted to the cloud server where it stores and exchanges the information with other edge devices and the main host server. This information is accessed by the main server to process these data in a machine learning scheme where it produces a set of forecast values of the data. Hence, this information can be used for threshold classification to predict and inform the engineers when there is an error during the printing process.

314a On the Systematic Use of Metabolomics for the Development of Large-Scale Kinetics for Industrial Bioreactors

Antonis Kokossis, Konstantinos Mexis, Stefanos Xenios

The paper addresses major challenges in the development of Digital Twins for Design-Build-Test-Learning (DBTL) cycles as they launched to produce biocatalysts for industrial biotechnology applications. Conventional models continue to regress century-old models (e.g. Michaelis-Menten and Monod) that disregard biocatalyst dynamics and quite often fail to reliably scale-up bioreactors. Digital Twins are instead challenged to capture a wide range of heterogeneous data available from the bioreactor operation, metabolomics, and omics. Once systematically captured, such data can be used to produce custom-made large-scale kinetics with direct reference to the biocatalyst reactions. They can also be used to reverse-engineer decisions for the better design of biocatalysts.

This study proposes a systematic and generic approach, part of a collective effort (Biondustry 4.0) supported by IBISBA (https://www.ibisba.eu/), the European Infrastructure on Industrial Biotechnology that brings together researchers from process systems engineering, synthetic biology, and computer science. Our approach introduces a systematic and efficient method for finding biosynthetic pathways for metabolic engineering of organisms to produce valuable chemicals. The process of designing biosynthetic pathways involves extracting novel metabolic pathways from complex metabolic networks that integrate known biological processes and predicted biotransformations. The objective of pathway search tools is to generate biologically relevant metabolic pathways in an automated manner that are both informative and easily comprehensible. To facilitate this task, we propose constructing searchable graph representations of metabolic networks. These pathways will subsequently be employed to generate large-scale kinetic models to produce self-regulated digital twins for the small scale, autonomous and on-site industrial biotechnology reactors in the context of Bioindustry 4.0. The development of digital twins will support the autonomous operation of biochemical reactors coupled with process control and online optimization (closed systems) and process and strain engineering. The purpose would be to compute process efficiency and guide strain engineering to improve the biochemical process for better efficiency and performance, as well as for tailoring of product portfolios. AI technology would be explored to support advanced, model-based control and online process optimization and to connect process and strain engineering. All processes will be simulated using the SPSE bioreactors modelling software for multiple industrial use cases. By integrating multi-omics data and established frameworks, our approach constructs large-scale kinetic models that capture cellular dynamics and provide more flexibility during bioprocess optimization. These models can simulate cell behavior and describe experimental kinetics, making them a valuable tool for constructing dataspace communities of in-silico curated kinetic models for different metabolic systems. These dataspaces will bring together relevant data and frameworks to facilitate data pooling and sharing, enabling the development of more robust large-scale kinetic models for digital twins in industrial biotechnology applications. Digital twinning of cellular processes (translation, metabolism) will be developed to capture the key features of biocatalyst function and will be embedded and tested within the bioreactor process twin. By incorporating diverse data sources and following established guidelines, our aim is to improve the accuracy and reliability of the generated kinetic models and accelerate the development of robust strains with novel metabolic pathways. This process will produce in-silico curated kinetic data that can integrate different stages of the Design-Build-Test-Learn (DBTL) cycle, enabling the simultaneous design of biocatalysts and process engineering.

Engineering kinetics are currently based on regression studies based on experimental data with limited reference to the underlying reaction pathways. The production of large scale metabolic kinetic models is also hindered by the uncertainty of predicting the kinetic parameters and producing the iterative Design-Build-Test-Learn (DBTL) cycle.

Engineering kinetics are currently based on regression studies based on experimental data with limited reference to the underlying reaction pathways. The production of large scale metabolic kinetic models is also hindered by the uncertainty of predicting the kinetic parameters and producing physiologically relevant and robust kinetic models. Instead, the ORACLE framework offers an attractive environment to generate populations of large-scale (curated) dynamics and a platform for in-silico kinetic models to check for physiological relevance and stability. Such models are valuable in scale-up studies, in the design and optimization of bioreactors, also to manipulate pathways as required to increase yields and selectivity.

Using the ORACLE framework, we are able to generate large populations of kinetic models but the biggest percentage of them are not stable. This paper explains a systematic approach that combines deterministic methods, data analytics and machine learning to accelerate realizable kinetics that could be set a basis to connect the dynamics of the cell with the dynamics of the process. Due to the small percentage of stable models, we deployed several ML methods (black-box models) trained on the saturation and thermodynamic displacement parameters of curated kinetic models to predict whether the inferred kinetic model is stable or not. Then, a CART decision tree (surrogate model) was trained to approximate predictions of the black box model and to extract rules that constrain the kinetic parameters of some critical enzymes, reducing the initial sampling space of the ORACLE framework and, as a result, the uncertainty in the model analysis.

The work is demonstrated with the production of muconic from S. cerevisiae which is achieved by shunting the shikimic pathway. The ORACLE framework consists of two reduction stages where the first one ensures that the generated kinetic models are physiologically relevant and the second one checks the model's stability. The conventional method produced 370 physiologically relevant models out of which 70 were stable (19.2%) whereas our approach increased uptake of acceptable solutions to 97.7%.

The uncertainty in the model analysis is reduced through the use of machine learning principles. Using machine learning classification and explainability techniques we were able to raise the stability index of the generated models which leads to postulate that it is better to constrain such systems as much as possible to get more feasible results.

Acknowledgments

The authors acknowledge support from the EPFL team of Prof. Hatzimanikatis,especially Miskovic Ljubisa, who assisted in applying the ORACLE methodology. Financial support from the project DEBONAIR (K.A. 61511600) funded by the Foundation for Research & Innovation is gratefully appreciated.

References

[1] Miskovic L, Hatzimanikatis V. Production of biofuels and biochemicals: in need of an ORACLE. Trends Biotechnol. 2010 Aug;28(8):391-7

[2] Xenios, S., Weilandt, D., Vasilis, H., Miskovic, L., & Kokosis, A. (2022). On the integration of process engineering with metabolomics for the production of muconic acid: the case for Saccharomyces Cerevisiae. Computer Aided Chemical Engineering, 541–546

[3] Andreozzi, S., Miskovic, L., & Hatzimanikatis, V., 2016, iSCHRUNK – In Silico Approach to Characterization and Reduction of Uncertainty in the Kinetic Models of Genome-scale Metabolic Networks. Metabolic Engineering, 33, 158–168.

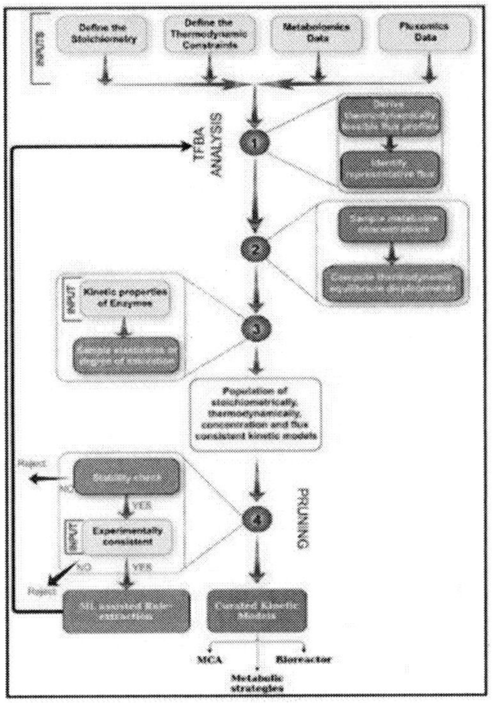

INPUTS
Define the Stoichiometry
Define the Thermodynamic Constraints
Metabolomics Data
Fluxomics Data
TFBA ANALYSIS
1
2
3
INPUT
Kinetic properties of Enzymes
Population of stoichiometrically, thermodynamically, concentration and flux consistent kinetic models
4
PRUNING
Reject
NO
YES
INPUT
Experimentally consistent
NO
YES
Reject
MCA
Bioreactor
Metabolic strategies

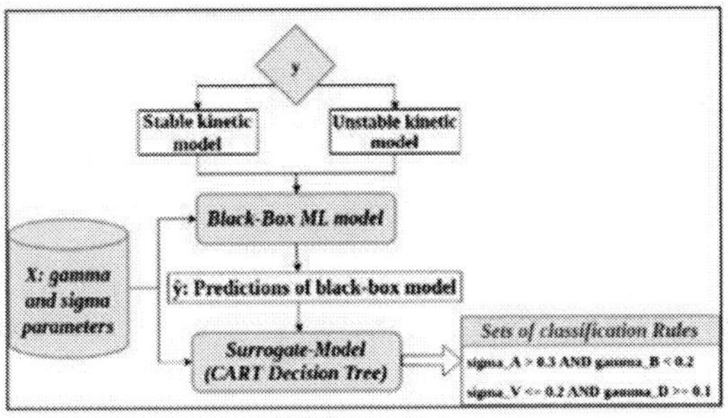

y
Stable kinetic model
Unstable kinetic model
Black-Box ML model
X: gamma and sigma parameters
ŷ: Predictions of black-box model
Surrogate-Model (CART Decision Tree)
Sets of classification Rules
sigma_A > 0.3 AND gamma_B < 0.2
sigma_V <= 0.2 AND gamma_D >= 0.1

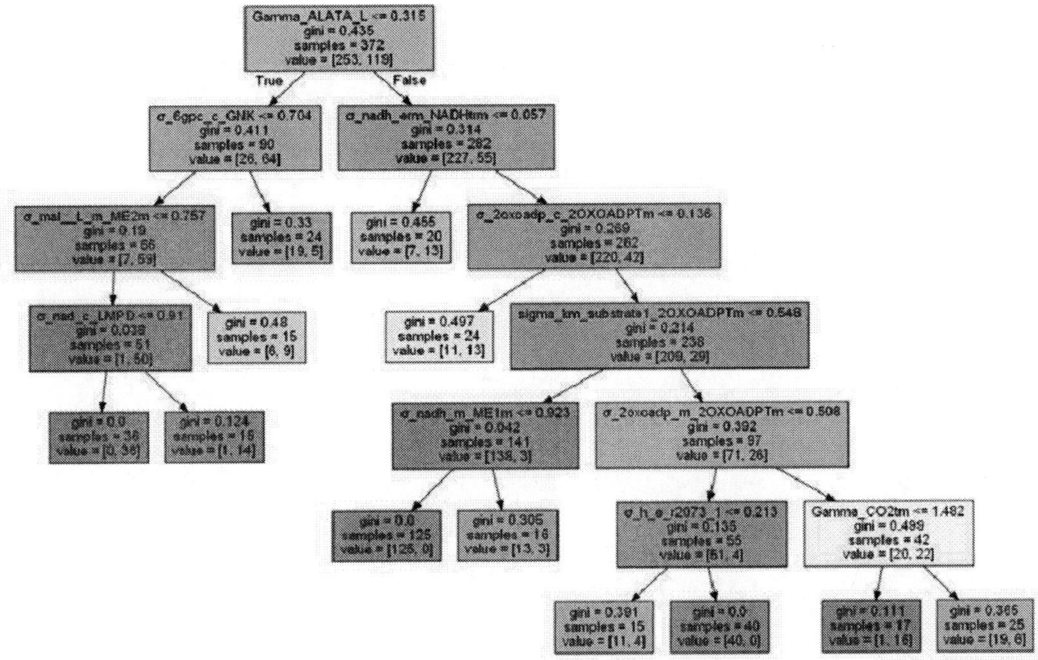

Gamma_ALATA_L <= 0.315
gini = 0.435
samples = 372
value = [253, 119]
True
False
σ_6gpc_c_GNK <= 0.704
gini = 0.411
samples = 90
value = [26, 64]
σ_nadh_em_NADHtm <= 0.057
gini = 0.314
samples = 282
value = [227, 55]
σ_mal__L_m_ME2m <= 0.757
gini = 0.19
samples = 66
value = [7, 59]
gini = 0.33
samples = 24
value = [19, 5]
gini = 0.455
samples = 20
value = [7, 13]
σ_2oxoadp_c_2OXOADPTm <= 0.136
gini = 0.269
samples = 262
value = [220, 42]
σ_nad_c_LNPD <= 0.91
gini = 0.038
samples = 51
value = [1, 50]
gini = 0.48
samples = 15
value = [6, 9]
gini = 0.497
samples = 24
value = [11, 13]
sigma_km_substrate1_2OXOADPTm <= 0.548
gini = 0.214
samples = 238
value = [209, 29]
gini = 0.0
samples = 36
value = [0, 36]
gini = 0.124
samples = 15
value = [1, 14]
σ_nadh_m_ME1m <= 0.923
gini = 0.042
samples = 141
value = [138, 3]
σ_2oxoadp_m_2OXOADPTm <= 0.508
gini = 0.392
samples = 97
value = [71, 26]
gini = 0.0
samples = 125
value = [125, 0]
gini = 0.305
samples = 16
value = [13, 3]
σ_h_e_r2073_1 <= 0.213
gini = 0.135
samples = 55
value = [51, 4]
Gamma_CO2tm <= 1.482
gini = 0.499
samples = 42
value = [20, 22]
gini = 0.391
samples = 15
value = [11, 4]
gini = 0.0
samples = 40
value = [40, 0]
gini = 0.111
samples = 17
value = [1, 16]
gini = 0.365
samples = 25
value = [19, 6]

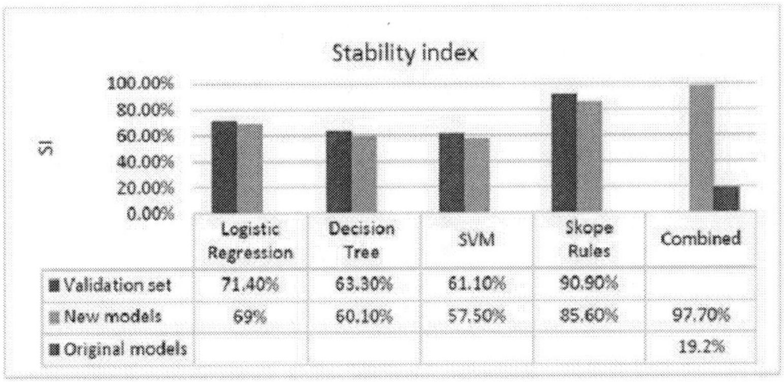

Stability index
SI
100.00%
80.00%
60.00%
40.00%
20.00%
0.00%
Logistic Regression
Decision Tree
SVM
Skope Rules
Combined
Validation set 71.40% 63.30% 61.10% 90.90%
New models 69% 60.10% 57.50% 85.60% 97.70%
Original models 19.2%

314b Digitalization of an Experimental Proton Conducting Membrane Reactor with Smart Manufacturing Principles

Berkay Citmaci, Derek Richard, Junwei Luo, Joonbaek Jang, Carlos Morales-Guio, Panagiotis Christofides

Smart manufacturing (SM) is a practice used in industry to optimize manufacturing processes and improve efficiency through enhanced sensing and the application of advanced data analysis techniques. These same techniques can be applied at the lab scale to accelerate the extraction of underlying relationships from complex processes. As such, SM techniques were applied to an experimental setup for high pressure and temperature electrochemical steam methane reforming based on proton conducting BZCY ceramic tubes [1]. This setup includes a tubular membrane electrode assembly, steam systems, pressure sensing and regulation, gas chromatography for product quantification, temperature sensing, gas flow control, and control of the electrochemical reactor with a potentiostat [2]. The system is connected to and operated through a single LabVIEW user interface that organizes system automation. This work is inspired by and built on the digitalization efforts from our previous work about electrochemical CO_2 reduction setup [3], that have leveraged automation strategies and reusable SM profiles, such as the automated gas chromatograph to quickly adapt to the new system and further improve it.

The experimental setup interacts with the CESMII's Smart Manufacturing Innovation Platform (SMIP) for data transfer, real-time monitoring, reporting, safe data storage, and hierarchical organization of process devices on the platform. The tools on SMIP are used to automatically generate and store calculated variables, deploy models, automatically train models, and online/offline data processing. Data collection is based on the design of each experiment, and procedures are scheduled to run the system through a series of conditions automatically to produce voluminous datasets at predetermined process conditions with minimal user oversight. This data is also collected in a way that facilitated the training of machine learning models and the design of system control strategies. Additionally, image processing techniques are applied to extract data from mobile cameras, such as a smartphone, to read digital screens of analog sensors, and record valve positions of the experimental setup then automatically update this information on the SMIP. In short, this work demonstrated how our research group is leveraging advanced technologies and collaboration with CESMII SMIP to advance the development of smart manufacturing techniques for emerging technologies like proton conducting membrane reactors.

References:

[1] Malerød-Fjeld, Harald, et al. 'Thermo-electrochemical production of compressed hydrogen from methane with near-zero energy loss.' *Nature Energy* 2.12 (2017): 923-931.

[2] Richard, Derek Michael. Development and Testing of Two Lab-Scale Reactors for Electrified Steam Methane Reforming. University of California, Los Angeles, 2021.

[3] Çıtmacı B; Luo J; Jang J; Morales-Guio, C.G.; Christofides PD; Digitalization of an experimental electrochemical reactor via smart manufacturing innovation platform. Digital Chemical Engineering 5, 100050.

314c Industry 4.0: Digital Transformation of Advanced Continuous Pharmaceutical Manufacturing

Ravendra Singh

Currently, Industry 4.0 concepts are being applied to pharma industry to achieve Pharma 4.0 paradigm. Pharma 4.0 reduces the time and resources needed for continuous pharmaceutical manufacturing and also improves the product quality and production consistency. It has many advantages but also have bigger challenges on the applications of artificial intelligence (AI)/machine learning (ML), material traceability, optimization, advanced process control, cyber-physical security, and data management side because of the different levels of complexities involved [1-3]. The predictive capabilities and the quality of the pharmaceutical products can be improved significantly via employing the artificial intelligence and the advanced model predictive control (MPC) system if an appropriate cyber-physical security defense is in place.

In this work, seven components of industry 4.0 namely artificial intelligence (AI)/machine learning (ML), modelling, material traceability, optimization, advanced control, cyber-physical security, and data science have been developed and implemented into the continuous pharmaceutical manufacturing process.

Four machine learning (ML) models have been trained to predict the response of continuous pharmaceutical manufacturing process and the performance of these ML models has been compared. The investigated ML methods are long short term memory (LSTM), 1D convolution neural network (CNN), random forest (RF), and artificial neural network (ANN). The best performing ML model is then implemented into the continuous pharmaceutical tablet manufacturing process for real time prediction. A digital twin model of continuous pharmaceutical tablet manufacturing process has been also developed.

A systematic framework including the methods and tools have been developed for material traceability. A corresponding software tool has been also developed to automate the material traceability procedure. The heart of the material traceability toolbox is the RTD model. The applications of the developed methods and tools have been demonstrated for material traceability of continuous pharmaceutical manufacturing process.

A systematic framework including the methods and tools has been developed for dynamic optimization of the feeder refill strategies. The deviation of the outlet mass flow of the feeder from the targeted flow rate has been minimized to obtain the optimum value of the feeder refill parameters. The material properties also affect the refill strategy meaning that the feeder refill strategy need to be frequently optimized if there are any changes in the materials and plant. Therefore, the developed feeder model and dynamic optimization tool can save the time and recourses as well as can improve the product quality significantly.

An advanced model predictive control (MPC) system has been implemented in the continuous pharmaceutical manufacturing (CPM) pilot-plant. The CPP's and CQA's are controlled in real time. The critical control variables that have been controlled using model predictive control (MPC) system are drug concertation, powder level before tablet press, main and pre compression forces, tablet weight and hardness. A novel control strategy for powder level control in a chute placed in between blender and tablet press unit operation of continuous tablet manufacturing process has been developed, implemented and evaluated.

A systematic framework including the methods and tools have been also developed for proactive identification and mitigation of potential cyber-physical attack risk on continuous pharmaceutical manufacturing plant [2]. The cyber-physical security relevant software tools such as Snap 7, Wireshark, and Tripwire have been applied to CPM. A novel software tool named CPS (Cyber-Physical Security) has been developed for cyber-physical security of the continuous pharmaceutical manufacturing. The integrated commercially available and developed (in house) cyber-physical security tools have added an extra layer of security of our continuous pharmaceutical manufacturing pilot-plant for any unexpected attacks.

All the relevant data generated during continuous manufacturing has been systematically collected, stored and organized in a data hub (OSI PI) and cloud system as per industry 4.0 standard.

References

1. Bhaskar, A., Barros, F. N., Singh, R. (2017). Development and implementation of an advanced model predictive control system into continuous pharmaceutical tablet compaction process. International Journal of Pharmaceutics, 534 (1-2), 159-178.
2. Billups, M., Singh, R. (2018). Systematic framework for implementation of material traceability into continuous pharmaceutical tablet manufacturing process. Journal of Pharmaceutical Innovation. https://doi.org/10.1007/s12247-018-9362-9.
3. Singh, R. (2021). Cyber-Physical Security (CPS) Tool for Continuous Pharmaceutical Manufacturing Process. Pharma Focus Asia. Issue 44, 41-49. https://www.pharmafocusasia.com/information-technology/cyber-physical-security-tool.

314e Web Application for LEAPS2 – a Surrogate Selection Framework

Maaz Ahmad, Sushant Garud, Damien Lee, Iftekar Karimi

Digital twins have been playing a key role in revolutionizing industrial process operations by creating accurate virtual models of real processes and physical assets. High-fidelity, first-principles-based models may be desirable as they are based on theoretical understanding of the system, but developing them requires system knowledge, expertise, time, and computational resource. Machine-learning-based data-driven surrogate models can serve as computationally cheaper and reliable alternatives to high-fidelity models. However, numerous options for modeling techniques (learning algorithms) and associated functional forms can produce a vast list of potential surrogate models. The choice of the best surrogate model form is rarely straightforward. It often entails time-consuming, exhaustive enumeration of many different surrogate models to select the model with the best accuracy. Surrogate selection procedure may get further complicated due to factors such as a user's preference for low model complexity along with good accuracy, and presence of noise in the data.

With this motivation, we developed a meta-learning-based surrogate selection framework, LEAPS2 (Learning based Evolutionary Assistive Paradigm for Surrogate Selection) (Ahmad and Karimi, 2021; Garud et al., 2018), to automate the surrogate selection process for any given data set. LEAPS2 operates in three modes – train mode, use mode, and update mode (see figure below). In the train mode, LEAPS2 learns surrogate performances over a large bank of non-noisy and noisy data sets by correlating the performances to certain data attributes. In the use mode, LEAPS2 uses its acquired knowledge to estimate surrogate performances for any given data set, and recommend a few best surrogates. In the update mode, LEAPS2 updates its knowledge by learning surrogate performances for new data sets, thereby progressively enhancing its knowledge with time.

We have developed a web-based application of LEAPS2 using Streamlit that can be accessed from https://leaps2.streamlit.app. As Streamlit is an open-source application framework, our tool can be used freely without the need of any third-party licensed software. The LEAPS2 framework within our tool has been trained on more than 5000 multi-dimensional (upto 20-dimensions) non-noisy and noisy data sets. Its surrogate pool comprises of 36 distinct surrogate forms from which a few surrogates are recommended. The tool comprises of three tabs – "Recommend", "Evaluate / Use", and "Help". The "Recommend" tab is the location where the user can receive surrogate recommendations for modeling a given data set. The data set is to be uploaded as a comma-separated values (csv) file in a specified format. In case of noisy data, our tool queries some optional inputs from the user to prevent recommendation of overfitting surrogates. It either uses a user-specified estimate of noise variance in the response data, or a threshold on minimum mean squared error or maximum R^2 to avoid recommending surrogates that may potentially fit noise in the response data. In the absence of user response, our tool uses a limit of . Our tool also provides the user the flexibility to choose a desired surrogate performance metric based on which the surrogates should be recommended. The user can either select an error-based metric (mean squared error, mean absolute error, and coefficient of determination or R^2) or a composite accuracy-cum-complexity metric (Surrogate Quality Score or SQS) (Ahmad and Karimi, 2021). Error-based metrics favor surrogates with low prediction error, while SQS combines model accuracy and complexity into a single score to favor simple and accurate surrogates. Each time a user uploads a data set in the "Recommend" tab, our tool seeks the user's permission whether their data can be saved for updating LEAPS2 in the future. The "Evaluate / Use" tab of the tool allows the user to assess the prediction performance of the recommended surrogates by performing hold-out validation. The split fraction to divide the data set into train and test sets can be specified by the user. Moreover, in this tab, the user can download a copy of the trained surrogates as a pickle file. The pickle file can be imported in Python, and the trained model can be used for making future predictions at new data points. Additionally, our tool allows the user to directly upload a csv file comprising of the test points to obtain corresponding predictions from the trained surrogates, without the need to download the pickle file. Detailed instructions on key features of the tool, and its usage are included within the "Help" tab of the tool.

Thus far, we have used our tool for making surrogate recommendations for various real-world data sets. To validate the recommendation performance of our tool, we obtained its recommendations on all the data sets used to train LEAPS2. Our tool successfully recommended at least one of the three true best surrogates for more than 90% of the data sets, highlighting its ability to often recommend best surrogates. With timely improvements based on comments and feedback, we hope that our tool can serve as a valuable component for industrial digitalization.

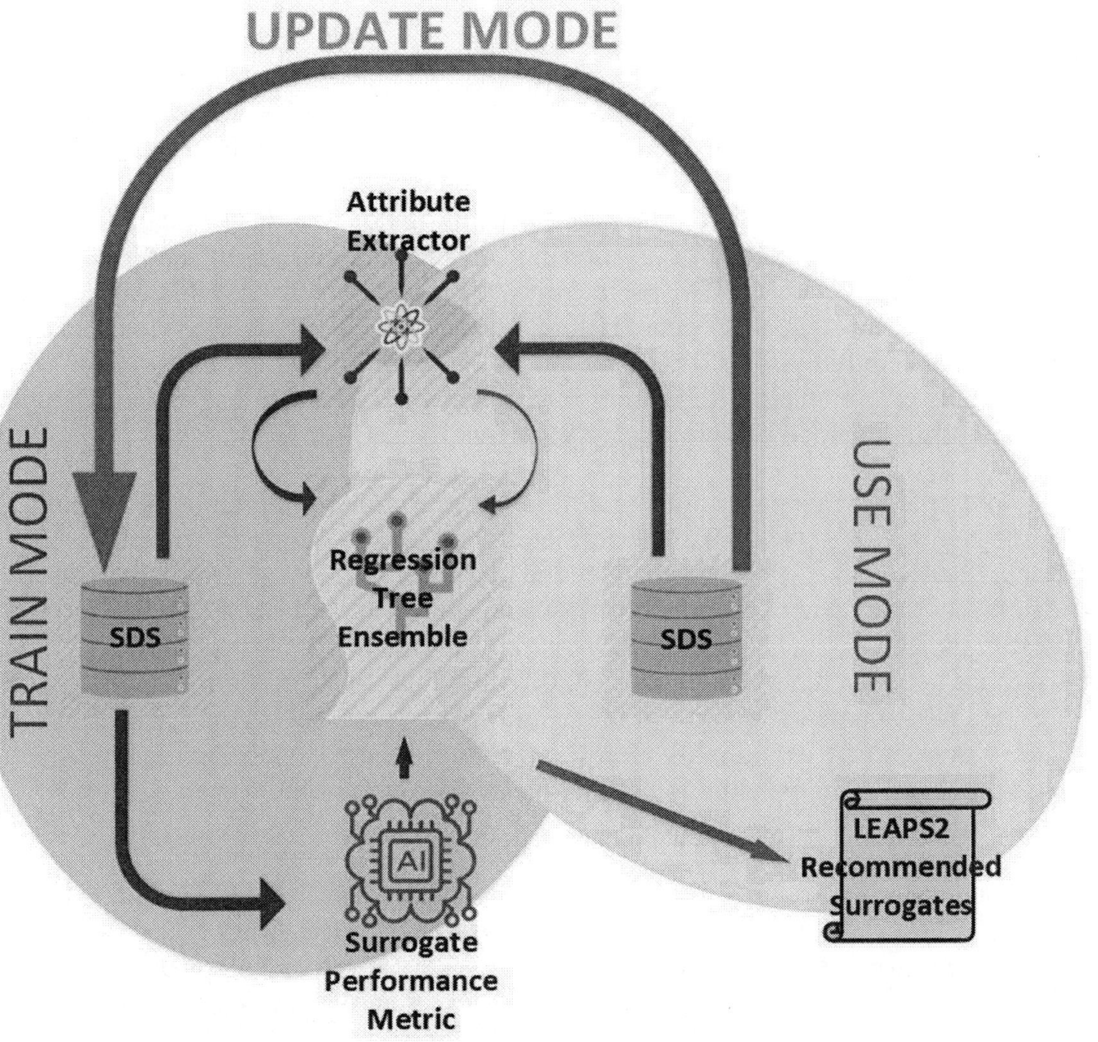

314f Statistical Evaluation of a Data Exception and Compression Algorithm Applied to Industrial Data Management Systems

Jose Edivaldo Junior Jr., Fernando Lima, Heleno Bispo

The integration and availability of data in the industry have played a crucial role in the operational and managerial activities of various industries over the years. However, with the intensification of data exploration through advanced statistical techniques and machine learning methodologies, data quality has become a critical priority (Xie et al., 2020). Such methodologies and techniques applied are predominant in industries that have reached or are moving towards the Industry 4.0 paradigm, highlighting the need for high-quality data that represents the real observed process behavior. Specifically, industrial data management is of paramount importance to industrial data infrastructure. Data historians play a crucial role in collecting, filtering, storing, and making available data collected from various sources (Thornhill et al., 2004; Alsmeyer, 2006). The objective of this work is the evaluation of the data filtering stage that occurs before data storing using the PI System infrastructure. In this infrastructure, data filtering is performed by the combination of two distinct algorithms (Bristol, 1990): the data exception algorithm, traditionally known as the deadband algorithm, and the data compression algorithm, named by Bristol (1987) as Swinging Door Trending.

The methodology presented in this paper aims to define the configuration parameters for both algorithms, quantifying the impacts of various parameter sets. As a tool for verification and comparison of data quality, some key performance indicators (KPI) have been implemented: (i) centrality indicators, such as Compression Ratio (CR) and Mean Squared Error (MSE); and (ii) variability centrality indicators, such as Ratio between the Variance of the Reconstructed Values (Ei), Percentage Difference between the Mean Values (PDM), and Pearson Correlation Coefficient (Rxy). Such KPIs were used to compare the quality of the resulting data with the original process data. With this, it is possible to evaluate the data quality after the filtration process, enabling the reproduction of the original data's process dynamics without attenuation or distortion of the variable profiles stored in the PI System Data Archive.

As a case study to showcase the framework, a multi-component hydrocarbon flash unit has been developed in AVEVA Process Simulation, with the flash vessel level dynamics considered under changing operating conditions. The data filtration process was explored by comparing the raw data to the data stored through PI Points in the PI System. By applying the methodology to different process intervals, appropriate parameters for the filtration process were estimated. Results on this method show that the simulation and validation of the data capture and filtration processes in the PI System have provided a reliable and effective approach for optimizing the compression and exception process parameters. In particular, it was possible to achieve an 80% reduction in the volume of the original data, while maintaining the original profile observed in the raw data and reproducing the process dynamics under consideration. The accumulated error was only 0.000638%, demonstrating the effectiveness of the data reduction approach.

Keywords: Industrial data infrastructure; Data historian; Data filtering; Process Dynamics.

References

Xie, H., Li, Y., Chen, G., Liu, Z., Chen, X. (2020), IEEE Transactions on Industrial Informatics, 16(5), 2964-2974

Bristol, E. H., (1987), US Patent 4,669,097.

Bristol, E. H., (1990), Proceedings of the ISA National Conference, 749–753.

Alsmeyer, F., (2006), Computer Aided Chemical Engineering, 21, 1533-1538.

Thornhill, N., Shoukat Choudhury, M., Shah, S., (2004), Journal of Process Control, 14(4), 389-398.

314g Exploring Model-Based Control Design for Next-Generation Etch Cooling Systems

Henrique Oyama, Dominic Messina, Kip Nieman, Helen Durand

Etch cooling systems for the semiconductor manufacturing industry are critical to maintaining wafer etching performance. In particular, precise control of the wafer temperature is required to meet industrial specifications, which involve features such as shape and sharpness at modern technology nodes [1]. As described in [1] and [2], the next-generation etching process will require modules that may be incorporated into modern wafer temperature control (WTC) systems to operate in a wider range of operating conditions, including cryogenic process modules. To achieve this, the selection and exploration of advanced control designs for etch cooling systems under different operating strategies must be carefully considered [3].

Advances in measurements and process control of wafer temperature units have been reported in the literature (e.g., [4], [5], [6], [7], [8], [9]). Particularly, in [7], fitting polynomial models for statistical quality control have been used for a statistical feedback control framework that targets quality features (e.g., etch rate). In [8], a model-based control approach for etch processes has been proposed to reduce the critical dimension variations on the domain of across-wafer levels. In [9], a recurrent neural network-based model predictive controller (MPC) for a plasma etch process has been developed considering a multi-scale model to simulate the gas flows and chemical reactions of plasma and the etching process on the substrate. The process models developed for testing the control methods above were reduced-order models based on experimental data. In light of this, a modular closed-loop digital model for WTC systems may be used for data generation and further analysis of operating strategies, which could allow the semiconductor manufacturing industry to explore advanced control strategies in etching systems that do not need to rely on the actual physical system for testing novel control schemes while still achieving a high level of confidence.

Motivated by the above, prior work in our group [10] developed a tunable ANSYS/CFD model that captures the major features of typical cooling etch systems. The closed-loop system under PID controllers has been selected for the proposed ANSYS/CFD model, which represents a control strategy typically implemented in the semiconductor manufacturing industry. However, this control strategy using PID controllers may not work well for modern etch cooling systems, especially considering that fast and large set-point changes could happen to achieve desired industrial conditions and features in the wafer. To tackle this challenge, we discuss the process of designing an MPC for the WTC system, including considerations such as the impact of the MPC prediction horizon on control performance and computational complexity, timescale differences between the states of the system (e.g., flow rate changes over the pipes happen in a much faster timescale than temperature changes), state measurement availability, and MPC optimization problem formulation. In particular, as an MPC framework for the WTC system has beneficial properties compared to the WTC under PID controllers, the performance of both control systems will be compared. Specifically, MPC is an optimization-based control design, meaning that it computes optimal control actions to implement in the WTC system while accounting for hard process constraints and making predictions of how the WTC system will behave under the MPC control actions in the future (characteristics for which are not included in the PID controllers). Thus, the exploration of MPC strategies for cooling etch systems can provide flexibility for optimization and may enable new types of operating conditions to increase wafer etching performance. To illustrate this, a closed-loop description of the digital model developed in [10] for typical cooling etch systems under an MPC framework is discussed.

References

1. Tachi, S., Tsujimoto, K., & Okudaira, S. (1988). Low-temperature reactive ion etching and microwave plasma etching of silicon. Applied physics letters, 52(8), 616-618.
2. Tinck, S., Tillocher, T., Georgieva, V., Dussart, R., Neyts, E., & Bogaerts, A. (2017). Concurrent effects of wafer temperature and oxygen fraction on cryogenic silicon etching with SF6/O2 plasmas. Plasma Processes and Polymers, 14(9), 1700018.
3. Lee, C. G., Kanarik, K. J., & Gottscho, R. A. (2014). The grand challenges of plasma etching: a manufacturing perspective. Journal of Physics D: Applied Physics, 47(27), 273001.
4. Sun, M., & Gabriel, C. (2002). Direct wafer temperature measurements for etch chamber diagnostics and process control. In 13th Annual IEEE/SEMI Advanced Semiconductor Manufacturing Conference. Advancing the Science and Technology of Semiconductor Manufacturing. ASMC 2002 (pp. 134-139). IEEE.
5. Wang, S. Q., Macdonald, P., & Kruger, M. (2006, October). Plasma Etch Process Diagnosis and Control By Wireless Sensor Wafer in Semiconductor Chip Manufacturing. In 2006 8th International Conference on Solid-State and Integrated Circuit Technology Proceedings (pp. 2175-2180). IEEE.
6. Lynn, S. (2011). Virtual metrology for plasma etch processes (Doctoral dissertation, National University of Ireland Maynooth).
7. Mozumder, P. K., & Barna, G. G. (1994). Statistical feedback control of a plasma etch process. IEEE Transactions on Semiconductor Manufacturing, 7(1), 1-11.
8. Zhang, Q., Poolla, K., & Spanos, C. J. (2008). One step forward from run-to-run critical dimension control: Across-wafer level critical dimension control through lithography and etch process. Journal of Process Control, 18(10), 937-945.
9. Xiao, T., Wu, Z., Christofides, P. D., Armaou, A., & Ni, D. (2021). Recurrent neural-network-based model predictive control of a plasma etch process. Industrial & Engineering Chemistry Research, 61(1), 638-652.
10. Oyama, H., Nieman, K., Tran, A., Keville, B., Wu, Y., & Durand, H. (submitted). Computational Fluid Dynamics Modeling of a Wafer Etch Temperature Control System. Digital Chemical Engineering.

377a A Digital Twin Model of Homogenizer Used in Continuous Injectables Manufacturing

Priya Das, Fernando Muzzio, Ravendra Singh

Currently, the pharmaceutical industry is going through a paradigm shift from batch to advanced continuous manufacturing (CM) of liquid dosages form. The continuous manufacturing provides several advantages including low footprint [1], better process control and safety [2], improved product quality [3], easier adaptation, lesser manufacturing and operating cost, and easier scale-up/number-up. However, the development, and adaptation of the continuous manufacturing of the liquid dosages forms is still a challenging task because of different levels of complexities. Therefore, systematic methods and digital tools are needed to support the quick development, adaptation, optimization, and control of the continuous manufacturing of the injectable liquid dosages forms.

Continuous homogenizer plays a very important role in the continuous manufacturing of liquid dosages forms. The products achieve better active ingredient dispersion after being processed through the homogenizer. This further helps to enhance the bioavailability and drug release profiles. Reducing the agglomerates &/or particles size and improving the uniformity in composition can also have a significant impact on the consumer appeal of the products, besides having other advantages. In this research, an advanced digital twin model of continuous homogenizer has been developed. This model describes the particle size reduction within the homogenizer which results in the generation of fines. Mechanical forces such as shear, turbulence, impact, cavitation, and centrifugal action are the primary causes, leading to the subdivision of particles into very small sizes for further processing. The action of homogenization leads to better dispersion of the API, thereby resulting in more stable products. It provides a clear understanding of the flow structure and the flow field parameters inside the homogenizer which is essential for the design and optimization of this unit operation. Being highly adaptive, it enables digital design, optimization and control of the homogenizer. The model has been validated for homogenization of ingredients used in albuterol production.

The applications of the model have been demonstrated for scenario analysis, sensitivity analysis and to optimize the process. Since mixing and particle size reduction are the key processes in any continuous manufacturing line, this study is essential for enhancing the overall efficiency and effectiveness of the process and the product.

References

1. Rogers, L., Jensen, K. (2019). Continuous manufacturing – the Green Chemistry promise? Royal Society of Chemistry, 21, 3481-3498.
2. Mascia, S., Heider, P., Zhang, H., Lakerveld, R., Benyahia, B., Barton, P., Braatz, R., Cooney, C., Evans, J., Jamison, T., Jensen,K., Myerson, A., Trout, B. (2013). End-to-End Continuous Manufacturing of Pharmaceuticals: Integrated Synthesis, Purification, and Final Dosage Formation. Angewandte Chemie, 125, 12585-12589.
3. May,S. (2017). Flow Chemistry, Continuous Processing, and Continuous Manufacturing: A Pharmaceutical Perspective. Journal of Flow Chemistry, 7, 3-4.

Acknowledgements

This work is supported by the US Food and Drug Administration (FDA) under contract number 75F40122C00122.

377b Image Based Control Virtualization

Akkarakaran Leonard, Helen Durand

Image Based Control (IBC) has seen wide-ranging applications in the control of autonomous robots [1, 2]. In chemical engineering, IBC has been used to gather process information, such as the online monitoring of crystal size distributions [3, 4, 5], and optimal combustion practices [6]. Virtual test beds for simulating IBC can be useful in setting baseline parameters, developing ideal scenarios, as well as controlled introduction of noise into a target system. The use of virtual image test beds is common in robotics, for applications like drone flight control [7], manipulation of surgical robots [8], and spatial exploration with mobile devices [9]. The target process envisioned in this work is the manipulation of monomers suspended in free space that self-assemble, also known as Directed Self-Assembly (DSA).

For this work, Blender, a free and open source 3D modeling and animation software, is used to simulate and control an IBC controlled process. The work aims to simulate an ideal IBC example of a DSA process. The process in this example involves the manipulation of multiple nanorods in two dimensions to achieve a desired configuration. The state of the rods, their position and orientation in the medium, are captured by the virtual camera in the Blender environment. Hypothetical control is enacted by translating each rod along its axis at a fixed velocity. This desired configuration is defined in relative distance and orientation, and as such the state of the system is a compact formulation of the positions of individual rods. Since the motion of the rods in the medium will be largely directed by Brownian Dynamics, a stochastic model predictive controller (SMPC) is set up to derive optimal control.

[1] Rafik Mebarki and Vincenzo Lippiello. “Image-based control for aerial manipulation”. In: Asian Journal of Control 16.3 (2014), pp. 646–656.
[2] Christophe Collewet and Fran¸cois Chaumette. “A contour approach for image-based control on objects with complex shape”. In: Proceedings. 2000 IEEE/RSJ International Conference on Intelligent Robots and Systems (IROS 2000)(Cat. No. 00CH37113). Vol. 1. IEEE. 2000, pp. 751–756.
[3] Xue Z Wang, Kevin J Roberts, and Caiyun Ma. “Crystal growth measurement using 2D and 3D imaging and the perspectives for shape control”. In: Chemical Engineering Science 63.5 (2008), pp. 1173–1184.
[4] Sebastian Maaß et al. “Automated drop detection using image analysis for online particle size monitoring in multiphase systems”. In: Computers & chemical engineering 45 (2012), pp. 27–37.
[5] Paul A Larsen, James B Rawlings, and Nicola J Ferrier. “An algorithm for analyzing noisy, in situ images of high-aspect-ratio crystals to monitor particle size distribution”. In: Chemical Engineering Science 61.16 (2006), pp. 5236–5248.
[6] Junghui Chen et al. “Design of image-based control loops for industrial combustion processes”. In: Applied energy 94 (2012), pp. 13–21.
[7] Dongliang Zheng et al. “Image-based visual servoing of a quadrotor using virtual camera approach”. In: IEEE/ASME Transactions on Mechatronics 22.2 (2016), pp. 972–982.
[8] Brian C Becker et al. “Vision-based control of a handheld surgical micromanipulator with virtual fixtures”. In: IEEE Transactions on Robotics 29.3 (2013), pp. 674–683.
[9] Richard Werner Nelem Pazzi, Azzedine Boukerche, and Tingxue Huang. “Implementation, measurement, and analysis of an image-based virtual environment streaming protocol for wireless mobile devices”. In: IEEE Transactions on Instrumentation and Measurement 57.9 (2008), pp. 1894–1907.

377c Algorithmic Self-Optimization of Processes in Continuous Flow

Sebastian Soritz, Nico Nys, Heidrun Gruber-Woelfler

Finding the ideal operating conditions of a chemical process has traditionally demanded manual and iterative laboratory work. Researchers will usually apply the one-variable-at-a-time (OVAT) approach until presumably ideal conditions have been found. Recent literature however, describes the use of complex strategies for process optimization, such as the design of experiments (DoE) approach, as well as algorithms from the field of mathematical optimization.[1] Combining such optimization strategies with automated laboratory equipment has allowed for "self-optimizing" chemical systems, also called "self-driving labs"[2] or more generally, "closed-loop optimization". These systems aim to find the ideal set of parameters to optimize a desired process step outcome, such as the yield of a chemical reaction, or the particle size for a crystallization step.

The advantages of utilizing these self-optimizing systems, when compared to the OVAT approach, include the reduction of manual laboratory work for researchers, the ability to simultaneously optimize multiple parameters and outcomes at the same time, which saves time and resources, and the increased certainty to find the true global optimum of a process step by leveraging advanced mathematical strategies.

However, due to the highly specialized nature of chemical procedures, the implementation of automated closed-loop optimization systems is a challenging endeavor.[3] Additionally, describing a chemical process as an optimization problem in mathematical terms, and subsequently finding the right algorithm for the problem at hand, has proven to be challenging in and of itself.

In this project, we strive to address the difficulties associated with self-optimizing systems laid out above. In our first work, we have established a laboratory setup for the automation and optimization of a biocatalytic reaction in flow via DoE, utilizing 3D printing of ceramics and enzyme immobilization.[4] Following this, we have worked on the benchmarking of different optimization algorithms, with the inclusion of currently underutilized algorithms, using both theoretical test problems, as well as a Suzuki coupling reaction in continuous flow.[5] Finally, we are applying the principle of self-optimization to the continuous crystallization of active pharmaceutical ingredients (APIs), via the utilization of 3D printing, optical particle analysis and Bayesian optimization, which has, to our knowledge, not been done before.

Overall, we have been able to advance the state of knowledge in the field of self-optimization by developing new automated setups in continuous flow in the areas of biocatalysis, homogeneous catalysis, and continuous crystallization of APIs, as well as further the understanding of algorithm usage for real-life process optimization.

1. Mateos, C., Nieves-Remacha, M. J. & Rincón, J. A. *Reaction Chemistry and Engineering* **4**, 1536–1544 (2019).
2. Abolhasani, M. & Kumacheva, E. *Nat. Synth* 1–10 (2023).
3. Christensen, M. *et al. Chemical Science* **12**, 15473–15490 (2021).
4. Valotta, A. *et al. Journal of Flow Chemistry* **11**, 675–689 (2021).
5. Soritz, S., Moser, D. & Gruber-Wölfler, H. *Chemistry–Methods* **2**, (2022).

377d Multiscale CFD Modeling of an Area-Selective Atomic Layer Deposition Process: Application to Discrete Feeding Reactor Configuration Design

Matthew Tom, Henrik Wang, Feiyang Ou, Gerassimos Orkoulas, Panagiotis Christofides

Multiscale CFD Modeling of an Area-Selective Atomic Layer Deposition Process: Application to Discrete Feeding Reactor Configuration Design

Matthew Tom[1], Henrik Wang[1], Feiyang Ou[1], Gerassimos Orkoulas[3], and Panagiotis D. Christofides[1,2]

1. Department of Chemical and Biomolecular Engineering, University of California, Los Angeles, CA
2. Department of Electrical and Computer Engineering, University of California, Los Angeles, CA
3. Department of Chemical Engineering, Widener University, Chester, PA

In response to the downscaling of semiconductor feature sizes and the increasing density of transistors occupying a semiconducting wafer, fabrication methods have become more stringent in design specifications. In particular, wafers must be appropriately fabricated to enable nanopatterning and facilitate the bottom-up procedure through self-alignment for transistor stacking [1]. Various manufacturing procedures employ thin-layer deposition processes, which deposit thin films in a self-limiting manner. However, conventional deposition processes such as atomic layer deposition (ALD) lack selective deposition, which disrupts nanopatterning and increases the need for postprocessing steps including atomic layer etching (ALE). In response to aforementioned obstacle, area-selective atomic layer deposition (ASALD) processes have emerged to promote self-limiting and chemoselective deposition through the integration of surface protection layers that interfere with reagent surface adsorption on non-growth area surfaces [2]. Additionally, ultra, thin-film uniformity has been an emerging challenge for large-scale deposition processes, which prompted the development of discrete feeding reactors that better control the behavior of the fluid dynamics through semi-instantaneous adsorbate and purge gas pulses that are confined in discrete regions [3,4]. While laboratory paradigms of the processes have been established, optimization and development of ASALD and discrete feeding reactors have not been studied due to the difficulties of relating the atomistic-mesoscopic surface kinetics behavior of the reactions to the macroscopic fluid dynamics of reagents within the reactor. *In silico* multiscale modeling serves to better characterize the two phases by conjoining a kinetic Monte Carlo (kMC) simulation that describes the surface kinetics behavior with computational fluid dynamics (CFD) that numerically evaluates the spatiotemporal behavior of the adsorbates and byproducts within the reactor. Additionally, multiscale modeling can generate large data sets without requiring substantial computational power and capital for raw materials and experimental design.

This work will establish an *in silico* multiscale modeling framework for simulating area-selective atomic layer deposition of a SiO_2/Al_2O_3 wafer through a discrete feeding reactor by conjoining a mesoscopic simulation using the kMC algorithm and a macroscopic simulation using CFD. First, a discrete feeding reactor configuration (which includes a wafer substrate) will be constructed through a mesh discretization procedure that is appropriately optimized for numerical accuracy and computational efficiency. Next, the CFD simulation will be customized to the ASALD process environment by defining various boundary, operating, and numerical method conditions as well as user-defined functions (UDFs), which will be employed to carry out the kMC simulation through each nodal element on the wafer surface. Lastly, the CFD simulation will be conducted for various operating and reactor design conditions (e.g., reagent flow rates and composition and number of discrete feeding ports) to establish optimal reactor configurations and operating conditions to enable process scale-up.

References:

- Li, M.-Y., Su, S.-K., Wong, H.-S. P., & Li, L.-J., 2019. "How 2D semiconductors could extend Moore's law," *Nature*, 567, 169-170.
- Mackus, A. J. M., Merkx, M. J. M., & Kessels, W. M. M., 2019. "From the bottom-up: toward area-selective atomic layer deposition with high selectivity," *Chemistry of Materials*, 31, 2-12.
- Kim, D. H., Park, J. C., Park, J., Cho, D.-Y., Kim, W.-H., Shong, B., Ahn, J.-H., & Park, T. J., 2021. "Wafer-scale growth of a MoS_2 monolayer via one cycle of atomic layer deposition: an adsorbate control method," *Chemistry of Materials*, 33, 4099-4105.
- Han, J. W., Jin, H. S., Kim, Y. J., Heo, J. S., Kim, W.-H., Ahn, J.-H., Kim, J. H., & Park, T. J., 2022. "Advanced atomic layer deposition: ultrathin and continuous metal thin film growth and work function control using the discrete feeding method," *Nano Letters*, 22, 4589-4595.

377e Transfer Learning-Based Modeling and Predictive Control of Nonlinear Process Network

Ming Xiao, Cheng Hu, Zhe Wu

Recurrent neural networks (RNNs) have been widely used in modeling nonlinear dynamic systems using sequential or time-series data [1]. However, in practice, it can be difficult and expensive to collect sufficient data for some real-world systems to build RNN models. Transfer learning that transfers knowledge from a source process with a large amount of data to a target process with very limited data provides an effective tool to overcome data scarcity in modeling chemical processes [2] and makes knowledge easily transferable to new processes [3]. Despite the increasing popularity and application of transfer learning in many fields of process systems engineering [4], the application of transfer learning in process modeling is still in its infancy. Specifically, the applicability and benefits of transfer learning for modeling nonlinear processes through the transfer of knowledge from a related task remain an open question. Furthermore, the traditional transfer learning method focuses on the modeling of a single process, and is challenging to be generalized to a process network with multistage processes, since it requires a similar process network as the source domain, which is difficult to obtain in practice. Incorporating domain knowledge into transfer learning is a potential solution, and the research in this area is still in its infancy.

In this work, a framework of transfer learning for modeling a target nonlinear process is developed by taking advantage of the knowledge learned in a different but similar source process. Specifically, transfer learning uses a pre-trained model developed based on a source domain as the starting point, and adapts the model to a target process with similar configurations. The generalization error for TL-based RNN (TL-RNN) is first derived to demonstrate the generalization capability on the target process. The theoretical error bound that depends on model capacity and the discrepancy between source and target domains is then utilized to guide the development of pre-trained models for improved model transferability. Subsequently, the TL-RNN model is utilized as the prediction model in model predictive controller for the target process. A simulation study of chemical reactors via Aspen Plus Dynamics is used to demonstrate the benefits of transfer learning Furthermore, to generalize the transfer learning method to a nonlinear process network, we integrate transfer learning with physics-informed machine learning methods to improve the overall prediction performance of the entire process network by incorporating domain knowledge such as process structural knowledge and the first-principles model into the training process. An example of multistage processes is used to illustrate the effectiveness of the proposed transfer learning strategy for chemical process networks.

References:

[1] Ren, Y. M., Alhajeri, M. S., Luo, J., Chen, S., Abdullah, F., Wu, Z., & Christofides, P. D. (2022). A tutorial review of neural network modeling approaches for model predictive control. Computers & Chemical Engineering, 107956.

[2] Jiang, J., Shu, Y., Wang, J., & Long, M. (2022). Transferability in deep learning: A survey. arXiv preprint arXiv:2201.05867.

[3] Neyshabur, B., Sedghi, H., & Zhang, C. (2020). What is being transferred in transfer learning?. Advances in neural information processing systems, 33, 512-523.

[4] Zhuang, F., Qi, Z., Duan, K., Xi, D., Zhu, Y., Zhu, H., Xiong, H. & He, Q. (2020). A

comprehensive survey on transfer learning. Proceedings of the IEEE, 109(1), 43-76.

[5] Zhang, Y., Liu, T., Long, M., & Jordan, M. (2019, May). Bridging theory and algorithm for domain adaptation. In International conference on machine learning (pp. 7404-7413). PMLR.

[6] Karniadakis, G. E., Kevrekidis, I. G., Lu, L., Perdikaris, P., Wang, S., & Yang, L. (2021). Physics-informed machine learning. Nature Reviews Physics, 3(6), 422-440.

377f The Pore Space of Packed Bed Reactors: From Characterization to Improvements

Anna Geohagan, Agnieszka Truszkowska

Packed bed reactors have been commonly used in the continuous synthesis of chemicals for many decades now [1, 2]. While established, these systems suffer from severe limitations obstructing their overall efficiency. The most notable issues are mass and heat transfer inefficiency, both leading to reduced yields of the desired product, high-pressure drops, and intense heat transfer with the cooling media required for the reactor to achieve optimal operation. Improvements in yield and energy uptake in these systems would have a substantial impact on both economy and sustainability as packed bed reactors are the systems behind some of the most massive chemical processes in the World, such as ammonia synthesis [3]. There are, unsurprisingly, many ongoing efforts to intensify and modernize these reactors [4].

Computational simulations represent powerful means to study packed bed reactors, devise, and test strategies for their improvements [5, 6]. A particularly important class of such reactors consists of randomly distributed porous pellets loaded with catalysts [1-4]. The spatial distribution of these pellets and their own pore structure are detrimental to the heat and mass transfer across the reactor. Thus, knowledge of the exact structure of the bed would benefit both computational and experimental researchers. While many models successfully simulate these systems without this information, pore-scale models have their unique advantages and capability to contribute to these ongoing improvement efforts [6]. At the same time, obtaining detailed information on the internal structure of a packed bed, and any porous material, requires the use of costly and not readily available techniques [6, 7].

In this work, we propose a method of determining the internal structure of a packed bed from inexpensive and easily obtainable laboratory and real-time process metrics. In particular, we use the average, bulk void fraction, hydraulic permeability, known pellet size, inlet and outlet feed temperatures, and product yields to obtain a statistically accurate description of the reactor structure. We use a combination of direct numerical simulations and an optimization algorithm to obtain an approximation of the reactor with spherical pellets that gives us the distribution of pore sizes, pore conductivities, and the fraction of open vs. closed pores. Equipped with this knowledge, we demonstrate potential avenues for process improvement that would not be accessible without a suitable representation of the pore space.

[1] Andrigo, P., Bagatin, R. and Pagani, G., 1999. Fixed bed reactors. Catalysis Today, 52(2-3), pp.197-221.

[2] Güttel, R. and Turek, T., 2016. Improvement of Fischer–Tropsch synthesis through structuring on different scales. Energy Technology, 4(1), pp.44-54.

[3] Zhou, D., Zhou, R., Zhou, R., Liu, B., Zhang, T., Xian, Y., Cullen, P.J., Lu, X. and Ostrikov, K.K., 2021. Sustainable ammonia production by non-thermal plasmas: Status, mechanisms, and opportunities. *Chemical Engineering Journal*, *421*, p.129544.

[4] Haase, S., Tolvanen, P. and Russo, V., 2022. Process intensification in chemical reaction engineering. Processes, 10(1), p.99.

[5] Stegehake, C., Riese, J. and Grünewald, M., 2019. Modeling and Validating Fixed-Bed Reactors:

A State-of-the-Art Review. ChemBioEng Reviews, 6(2), pp.28-44.

[6] Jurtz, N., Kraume, M. and Wehinger, G.D., 2019. Advances in fixed-bed reactor modeling using particle-resolved computational fluid dynamics (CFD). Reviews in Chemical Engineering, 35(2), pp.139-190.

[7] Chen, L., He, A., Zhao, J., Kang, Q., Li, Z.Y., Carmeliet, J., Shikazono, N. and Tao, W.Q., 2022. Pore-scale modeling of complex transport phenomena in porous media. Progress in Energy and Combustion Science, 88, p.100968.

377g Fault-Prognostic Explicit Model Predictive Control with Physics-Data Driven Monitoring

Austin Braniff, Md Al Masud, Yuhe Tian

Process safety management is a key priority in the industrial setting to ensure a secure working environment for all. However, to date, process safety incidents still cause severe losses to finances over $1 billion per year, as well as pressing societal and environmental consequences [1]. Existing process safety assessment methods (e.g., hazard and operability study, quantitative risk analysis) typically evaluates a static picture of the process facility in a certain phase of plant design without considering the dynamically varying operating conditions under uncertainties or driven by real-time decision making which may result in process failures [2-3]. On the other hand, industrial control systems do not explicitly account for process safety performance when calculating for optimal control actions. More recently, several works incorporated operational safety considerations in model predictive control striving to maintain state variables in safe bounds [4-6]. However, there still lacks a systematic strategy which can integrate formal process safety metrics with advanced control for real-time monitoring and proactively reduction of failure probability while simultaneously addressing uncertainties.

To meet these needs, we present a framework for dynamic risk monitoring, online model updating, and fault-prognostic control as shown in Figure 1. A physics-based statistical dynamic risk model [7] is adapted to quantify the impact on real-time failure probability and consequence severity, as a result of safety-critical process variable deviations from nominal operating conditions. A multi-parametric model predictive controller (mp-MPC) is then designed based on the process model and dynamic risk model [8]. The multi-parametric control laws can thus be derived offline in priori as piecewise affine functions of process states, outputs, risk, and disturbances [9]. During online operation, the controller will determine optimal control actions by looking up the mp-MPC solution map instead of repetitively carrying out dynamic optimization. This risk-aware mp-MPC also offers the advantage of dual-layer process safety management by explicitly: (i) accounting for operating bounds of state variables and risk values in path constraints, and (ii) controlling risk as an output variable. The mp-MPC moving horizon policy also enables fault prognosis into the next output horizon leveraging model-based risk propagation forecast, which will in turn affect the optimal control decision at the current time step to proactively avoid or mitigate fault probability. If however, a critical failure is predicted and there is no action that can avoid it, alarm systems can relay this so that emergency protocols begin well before failure occurs. Moreover, an online model updating system is developed and coupled with the control system which takes advantage of artificial neural network to historicize process data and use it to detect and correct any modeling uncertainties. We will also showcase the online model updating to address model approximation errors if linear mp-MPC is preferred. The potential and efficacy of this framework is demonstrated with a batch reactor case study of the incident that occurred at T2 laboratories in Florida, USA Dec. 2007. [10]

References

1. Marsh, G. C., Mercer, O. W. & Companies, M. The 100 Largest Losses 1978-2017 Large property damage losses in the hydrocarbon industry. Marsh Report (2018).
2. Lee, J., Cameron, I., & Hassall, M. (2019). Improving process safety: What roles for Digitalization and Industry 4.0?. Process safety and environmental protection, 132, 325-339.
3. Villa, V., Paltrinieri, N., Khan, F. & Cozzani, V. Towards dynamic risk analysis: A review of the risk assessment approach and its limitations in the chemical process industry. Safety

Science 89, 77–93 (2016).
4. Leveson, N. G. & Stephanopoulos, G. A system-theoretic, control-inspired view and approach to process safety. AIChE Journal 60(1) (2014).
5. Albalawi, F., Durand, H. & Christofides, P. D. Process operational safety via model predictive control: Recent results and future research directions. Computers & Chemical Engineering 114, 171–190 (2018).
6. Venkidasalapathy, J. A. & Kravaris, C. Safety-centered process control design based on dynamic safe set. Journal of Loss Prevention in the Process Industries 65, 104126 (2020).
7. Bao, H., Khan, F., Iqbal, T., & Chang, Y. (2011). Risk-based fault diagnosis and safety management for process systems. Process Safety Progress, 30(1), 6-17.
8. Pistikopoulos, E. N., Diangelakis, N. A., Oberdieck, R., Papathanasiou, M. M., Nascu, I., & Sun, M. PAROC – An integrated framework and software platform for the optimisation and advanced model-based control of process systems. Chemical Engineering Science, 136, 115-138 (2015).
9. Ali, M., Cai, X., Khan, F., Pistikopoulos, E. N., & Tian, Y. Dynamic Risk-based Process Design and Operational Optimization via Multi-Parametric Programming. Under Review.
10. Chemical Safety Board, T2 laboratories inc. reactive chemical explosion. https://www.csb.gov/t2-laboratories-inc-reactive-chemical-explosion/

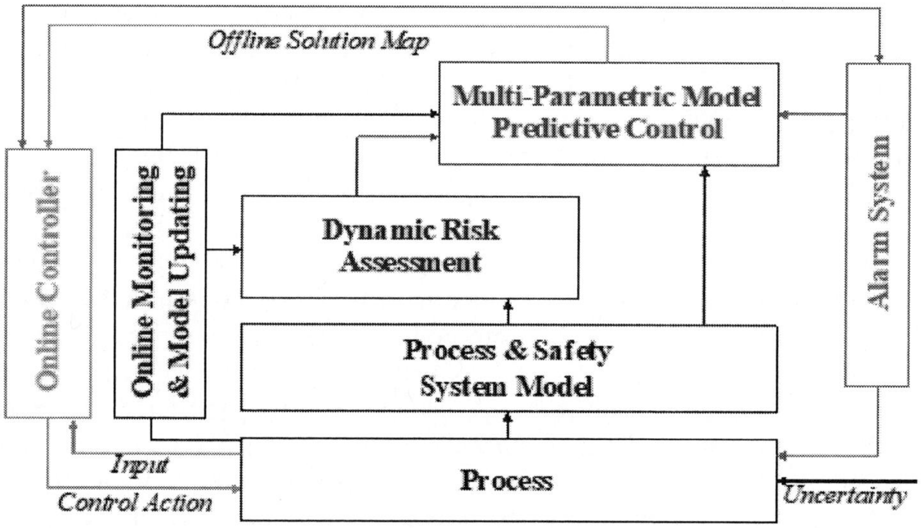

Figure 1. An overview of the proposed framework.

429a An Efficient, Cost-Effective, Continuous Polymer Purification Method

Alex Attard, Qing Pu, Zhiqiang Fan, Baxton Owen, Neal LaBanca, Walter Barnes

Polymer precipitation is a crucial process in chemical engineering and is widely applicable to many industries including semiconductors, pharmaceuticals, and wastewater treatment. In many industries, batch polymer precipitation is a common technique for polymer purification. However, batch polymer precipitation is an arduous and inefficient process, which generates excess solvent waste. It is labor-intensive, time-consuming, and demanding. We have designed a more optimal method to purify polymers by removing residual monomers and impurities from polymer solutions using continuous polymer precipitation. Continuous polymer precipitation allows for ease in scale-up of polymer purification and would provide significant energy and cost savings. Furthermore, while batch polymer precipitation requires several additional steps to get back to the same original polymer solution, this novel continuous process eliminates several steps and significantly decreases processing time. Thus, our continuous system is efficient, significantly reduces impurities in polymer solutions, and provides a viable, cost-effective alternative to batch polymer precipitation.

Continuous static mixing processes have been explored in the pharmaceutical industry relating to precipitation of nanoparticles[1]. However, this variation has not been explored in depth in the semiconductor industry. In the proposed process, after the polymer chains collapse, aggregate, and precipitate, the liquid waste is continuously isolated from the solid polymer and the solid polymer is completely dried. The purified polymer is then re-diluted to the same conditions of the original polymer solution. We have optimized this continuous process using a Design of Experiments (DOE) approach by varying factors such as the ratio of antisolvent to polymer solution, the flow rate of the polymer solution, and the retention rating of the phase-separation component. Additionally, computational fluid dynamics (CFD) were utilized to complete optimization of this process to improve the yield and efficiency.

The molecular weight (M_w) and high-performance liquid chromatography (HPLC) profiles show a significant reduction in residual monomer concentrations. This provides a significant advantage, as these impurities often have negative impacts on the properties and performance of the final product, including sublimation and defects. The antisolvent required for this continuous process has been reduced by 37% - 52% when compared to a batch process. The expected sublimation defects for the final product have been reduced by 53% when compared to the original polymer solution. Furthermore, the residual monomers have been significantly reduced through this continuous precipitation process, ranging from 67% - 86% reduction. For high-demand manufacturing materials, this continuous process would reduce polymer purification processing time by over 63% when compared to batch polymer precipitation. CFD has been used to optimize the location and number of mixers in addition to the ratio of antisolvent to polymer solution, which allowed us to maximize the yield of this process. The reduction in residual monomers and impurities, reduction in processing time, and system design optimization through CFD clearly demonstrate the advantages of this continuous precipitation process over the traditional batch precipitation process. This continuous polymer precipitation system has the potential to revolutionize the way polymer solutions are purified, making the process more efficient, cost-effective, and sustainable.

1. Dong, Yuancai. "A continuous and highly effective static mixing process for antisolvent precipitation of nanoparticles of poorly water-soluble drugs". International Journal of Pharmaceutics 386 (2010) 256-261.

429b Artificial Intelligence-Guided Manufacturing of Paints/Coatings

Daniel Schwartz, Dominic V. Poerio, Michael C. Grady, Anthony Calabria, Ali Shokoufandeh, Masoud Soroush

Abstract

The size of the global paints and coatings market is forecasted to grow from US$184 Billion in 2021 to US$212 Billion by 2026, which is a compound annual growth rate of 2.9% over the five years [1]. In view of the growing market size, incremental improvement in the quality of these products or minor decrease in their production costs will translate into significant economic gains for the industry [2]. A modern paint/coating (P/C) product is a complex mixture of many interacting compatible chemicals, the properties and amounts of which and how the chemicals are mixed together, strongly affect the properties of the P/C product. The production of modern P/C products with desired properties is challenging due to several factors [2]. First, there are many product attributes (e.g., for a basecoat there are more than fifteen quality attributes). Second, the ingredients of P/C products should be blended and mixed very carefully. Third, our current understanding of the different processes that produce P/C products is of varying degree. Fourth, the relationships between (a) the quality attributes of a P/C product and (b) the properties and amounts of the chemicals that the product are made from and the process type and conditions are complex and difficult to fully describe using physicochemical models. As a result of these challenges, paints and coating manufacturers currently use designed experimentation to customize the P/C products, and this approach limits the ability to predict properties outside the experimental space studied [3].

Given the wealth of historical data most paint manufacturers have on hand from years of product development, we embarked on developing data-driven models that combine unsupervised parameter estimation and supervised classification and prediction to deal with engineering systems with massive amount of historic data. This work deals with capturing, describing, and predicting the relationships using machine learning (ML) models trained on formula data from Axalta Coating Systems' historical P/C data. It also proposes advances necessary in ML methods to better capture these relations. We have attempted to improve current techniques using multi-channel deep neural network models. Our motivation for using multi-channel models is to separate the representation and fusion of information driven by historic data from the specific tasks of classification and prediction. This work is the first reported investigation of the use multichannel machine learning models in paints/coating industries. We present the formulated ML problems and different ML methods that we applied. We then present and discuss the ML results. Such mathematical models will enable P/C manufacturers to design and validate customized P/C products in silico, and produce customized P/C products with consistent properties in the presence of operation uncertainties.

Keywords

Data-driven modeling, neural networks, polymer resins, paints and coatings, optimization, color matching

References

[1] MarketsAndMarkets. "Paints & Coatings Market by Resin (Acrylic, Alkyd, Epoxy, PU, Vinyl, Fluoropolymer, Polyester), Technology (Waterborne, Solventborne, Powder), Application (Architectural (Residential, Non-residential), Industrial) and Region - Global Forecast to 2026". In: 2022. url: %5Chttps://www.marketsandmarkets.com/Market-Reports/paint-coating-market-156661838.html.

[2] Masoud Soroush and Michael C Grady. "Smart manufacturing of paints and coatings". In: Smart Manufacturing. Elsevier, 2020, pp. 179–218.

[3] Dispersions&Resins. 2022. url: https://dispersions-resins.basf.com/global/en.html.

429c Machine Learning-Based Gas Product Estimation and Feedback Control of an Experimental Proton Membrane Reactor

Berkay Citmaci, Derek Richard, Junwei Luo, Joonbaek Jang, Carlos Morales-Guio, Panagiotis Christofides

Steam methane reforming (SMR) is the most widely used hydrogen generation method in the chemical manufacturing industry. However, SMR is very energy intensive and requires pressure swing adsorption to separate hydrogen from the side products. An alternative method using proton membrane reformers (PMR), which is an electrochemical process powered by renewable resources to produce hydrogen from methane and water vapor, is more environmentally friendly process than the traditional SMR in terms reducing emission. Additionally, the ceramic membrane allows electrochemical separation and compression of hydrogen from the reactants, thus enable controlling the selectivity for CO and CO_2 by shifting the reaction equilibrium [1]. However, scaling up this complex technology for industrial applications poses challenges, including a lack of comprehensive process control strategies. This study focuses on developing efficient process control strategies for proton membrane reformers (PMRs) to optimize their operation and enable widespread adoption for sustainable hydrogen production. To accomplish this, machine learning models and model predictive control (MPC) will be implemented to optimize PMR operation.

An experimental PMRs system is developed, automated and tested at UCLA [2]. Data generated from this system is then used to develop ML model, such as Physics Informed Neural Networks (PINN), recurrent (RNN), and feedforward neural networks (FNN) that is expected to comprehensively capture the process behavior. Controlling the PMR system will require a multi-output-multi-input (MIMO) control system. ML-based estimators will be incorporated into an MPC to drive the process to most profitable or energy optimal set points. Thus, a comprehensive optimization scheme will be established to adequately control the reactor performance. Model construction will be supported by steady state and dynamic experimental applications to ensure comprehensive system performance prediction and control, and results will be compared with traditional SMR.

References:

[1] Malerød-Fjeld, Harald, et al. 'Thermo-electrochemical production of compressed hydrogen from methane with near-zero energy loss.' Nature Energy 2.12 (2017): 923-931.

[2] Richard, Derek Michael. Development and Testing of Two Lab-Scale Reactors for Electrified Steam Methane Reforming. University of California, Los Angeles, 2021.

429d Exploring the Environmental Impacts and Scalability of Mxene Synthesis for Advanced Applications

Mostafa Dadashi Firouzjaei, Mark Elliott, Babak Anasori

The next generation of manufacturing in chemical and energy systems is challenged by the scaling up and environmental impacts associated with nanomaterial synthesis. MXenes, discovered in 2011, are groundbreaking nanostructures with over 100 predicted compositions and over 40 lab-scale fabricated structures. Their unique properties make them ideal for numerous applications, such as energy storage, environmental, electronics, communication, gas and liquid separation, adsorption, biomedical, and optoelectronics. By 2023, MXenes demonstrated potential in energy storage devices, electromagnetic interference shielding, nanocomposites, and hybrid materials while outperforming other nanomaterials in emerging applications, like tribology. Despite these advancements, the environmental impacts of MXene synthesis remain unexplored. This study investigates the environmental impacts of fabricating $Ti_3C_2T_x$ MXene as a coating layer for electromagnetic interface (EMI) shielding in communication satellites, comparing it to aluminum and copper alternatives. Two $Ti_3C_2T_x$ synthesis systems, small and large laboratory scales, are designed with location impacts considered. The use of TiO_2 as an alternative to Ti in the synthesis of MAX is also examined. Ultimately, this work aims to contribute to the development of more efficient and environmentally friendly MXene synthesis, which is crucial for scaling up MXene production in next-gen chemical and energy systems.

429e Development and Implementation of Customized Symbols Based on the Greenscope Methodology for Sustainability Evaluation of Industrial Processes in the PI System

Esmael Gadelha, Fernando Lima, Heleno Bispo

Industry 4.0 represents the fourth industrial revolution, which leverages advanced technologies such as the Internet of Things (IoT), artificial intelligence (AI), and big data analytics to optimize industrial processes. Unlike previous industrial revolutions, Industry 4.0 emphasizes the collection and analysis of vast amounts of real-time data, which enables precise and efficient decision-making (Kestering et al., 2023). The acquisition, automation, and digitalization of processes are crucial for organizations to stay competitive in this new era. One of the significant benefits of Industry 4.0 is the regulation of information flow throughout the production process. Real-time data collection enables the identification of deviations, implementation of modifications/customizations, and ensuring the quality of the final output. With the ability to monitor key variables and indicators, Statistical Process Control (SPC) is a powerful tool that can identify anomalies that may have an impact on product quality (Montgomery, 2016).

This work is a continuation of a study presented in AIChE 2022 (Bispo et al., 2022), in which key process indicators were implemented in the PI System, to be evaluated in customized symbols implemented in PI Vision. The objective now is to expand the set of indicators based on the GREENSCOPE methodology, which allows for the evaluation of the sustainability of production processes (Ruiz-Mercado et al., 2014). For this purpose, a sustainability scale has been proposed for each indicator, delimited by two scenarios: the best and the worst case. This nondimensional scale transforms indicator scores into discrete unidimensional values between the selected best target and worst-case scores, facilitating visualization and comparison of sustainability assessment results for each indicator in four areas (Ruiz-Mercado, 2011). The structure implemented in the PI System has allowed the monitoring of the sustainability scale using a customized radar plot, as well as online verification based on SPC techniques. The integration of the GREENSCOPE methodology with the monitoring and control tools of the PI System has allowed a comprehensive and integrated point of view on process sustainability, enabling more efficient and precise decision-making.

References

Kestering, D., Agbleze, S., Bispo, H., Lima, F. V. (2023), Model predictive control of power plant cycling using Industry 4.0 infrastructure, Digital Chemical Engineering, 7.

Montgomery, D. C. (2016). Introduction to Statistical Quality Control (7th ed.). John Wiley & Sons.

Ruiz-Mercado, G.; Smith, R. L.; Gonzalez (2014). GREENSCOPE: Technical User's Guide. U. S. Environmental Protection Agency.

Bispo, H., Gadelha, E., Lima, F. V., Dantas, B., Almeida, N. A. B., Guerra, I. (2022). Development and Application of Customized Symbols in PI Vision for KPI Monitoring Based on the Greenscope Methodology. Proceedings of the AIChE Annual Meeting 2022, vol. 118 (1), pp. 2345-2356. DOI: 10.1016/j.proche.2022.03.045

Ruiz-Mercado, G. J. (2011). Sustainability Indicators for Chemical Processes: I. Taxonomy. Industrial & Engineering Chemistry Research.

429f Model Predictive Control of an Experimental Electrochemical Reactor

Junwei Luo, Berkay Citmaci, Joonbaek Jang, Derek Richard, Fahim Abdullah, Carlos Morales-Guio, Panagiotis Christofides

The electrochemical reduction of carbon dioxide (CO_2) has gained significant attention recently as an alternative source of manufacturing carbon-based fuels attributed to the rise in clean energy supply [1, 2]. With CO_2 reduction becoming an increasingly viable option for producing chemicals in an environmentally friendly manner, it has become necessary to model, optimize, and control the electrochemical CO_2 reduction process. In order to better comprehend the impact of mass transfer and reaction kinetics on productivity, a rotating cylinder electrode (RCE) reactor was constructed at UCLA [3], which allows for the analysis of these factors independently. Specifically, the RCE reactor employs two manipulated control variables to separate the kinetic and mass transfer effects on the reaction, based on the physical-chemical phenomena where the applied potential strongly influences reaction energetics, while electrode rotation controls hydrodynamics and convective mass transfer. In our recent work, a multi-input multi-output (MIMO) control scheme is developed on the RCE reactor to control the production rate of ethylene and carbon monoxide, utilizing techniques of neural network modeling, nonlinear optimization, and proportional integral (PI) control [4]. However, the RCE system generates sixteen liquid/gas products, which outnumber the available control inputs. This implies that the PI control method may have limitations in achieving comprehensive control over the production rate of all product species.

This paper aims to develop a model predictive control (MPC) to implement the MIMO control scheme that is capable of handling the situation with an unbalanced number of control states. To achieve this target, the production rate of hydrogen is included as an additional output state of the control system, which means there are three outputs (i.e., production rate of CO, C_2H_4, and hydrogen) and two inputs (i.e., surface potential and rotation speed) in the MIMO system. Considering the computational efficiency of solving the optimization problem in the MPC, a quadratic programming based MPC is chosen to carry out the real-time control process. Lastly, the performance of the developed MPC will be supported by experimental evidence and evaluated in comparison to the previously developed PI control system.

References:

[1] Morales-Guio, C.G.; Cave, E.R.; Nitopi, S.A.; et al. Improved CO_2 reduction activity towards C2+ alcohols on a tandem gold on copper electrocatalyst. *Nat Catal* 1, 764–771 (2018).

[2] Jang, J.; Shen, K.; Morales-Guio, C.G.; Electrochemical direct partial oxidation of methane to methanol. *Joule*. 2019, 3(11), 2589-93.

[3] Morales-Guio, C.G.; Jang J; Rüscher M; Winzely M.; Gastight Rotating Cylinder Electrode: Towards Decoupling Mass Transport and Intrinsic Kinetics in Electrocatalysis. Authorea Preprints. 2021 Oct 18.

[4] Çıtmacı B; Luo J; Jang J; Morales-Guio, C.G.; Christofides PD; Machine learning-based ethylene and carbon monoxide estimation, real-time optimization, and multivariable feedback control of an experimental electrochemical reactor. Chemical Engineering Research and Design. 2023 Mar 1;191:658-81.

429g Applications of Artificial Intelligence in Chemical Processes with Special Application in Heat Exchangers

Emmanuel Dada, Mario Lumueno, Elizabeth Osadare, Sarhan Musa

Heat exchangers are used in refrigeration and air conditioning systems, automobiles, thermal power plants, chemical and textile processing industries, and other industrial applications. They are devices that allow effective heat transfer between two fluids that differ in temperature. Because of its shape and the physical events that occur during heat transfer between fluids, heat exchangers are difficult to study. Heat exchangers are studied theoretically and empirically using the first and second laws of thermodynamics [1]. Theoretical heat exchanger analysis necessitates more assumptions and complicated equations, whereas experimental heat exchanger analysis is more expensive owing to the initial expenditure in establishing an experimental equipment [2]. Heat transport is a complicated phenomenon that may be modeled using artificial intelligence. One of the artificial intelligence approaches that can give valuable tools for modeling and correlating actual heat transfer problems is artificial neural networks (ANN). Mohamedi et al. [3] utilized the ANN approach to forecast the transport and thermodynamic characteristics of water's saturated vapor and saturated liquid, which are also used to predict the transport and thermodynamic properties of water's saturated vapor and saturated liquid [4]. One of the major advantages of ANNs is their capacity to learn, generalize, or extract automatically rules from complicated data; nevertheless, the technique's main disadvantage is that it can only be utilized in the range in which it was trained because it is empirical in nature [5]. This study involves a comparison and contrast of a heat exchanger process using a programming code language and ASPEN Plus simulation to determine the results that will be used as a basis of the artificial intelligence approach on heat exchangers.

1. L. Cornelissen, G.G. Hirs, Thermodynamic optimization of a heat exchanger, Int. J. Heat Mass Transfer 42 (1999) 951-959.
2. K. Shah, D.P. Sekulic, Fundamentals of Heat Exchanger Design, John Wiley & Sons, New York, 2003.
3. Mohamedi, B., *et al.*, Simulation of Nucleate Boiling under ANSYS-FLUENT Code by Using RPI Model Coupling with Artificial Neural Networks, *Nuclear Science and Techniques, 26* (2015), 4, pp. 40601-040601
4. Ghalem, N., *et al.*, Prediction of Thermal Conductivity of Liquid and Vapor Refrigerants for Pure and Their Binary, Ternary Mixtures Using Artificial Neural Network, *Thermophysics and Aeromechanics, 26* (2019), 4, pp. 561-579
5. Haykin, S., *Neural Networks: A Comprehensive Foundation,* Prentice Hall PTR, Upper Saddle River, N. J., USA, 1998

EMMANUEL DADA

Dr. Dada currently serves as an Assistant Professor of Chemical Engineering at Prairie A & M University. His research interests include engineering design and optimization, chemical process and technologies, artificial intelligence, plastics waste, and clean energy.

SARHAN MUSA

Dr. Musa currently serves as Professor of the Electrical Engineering Department at Prairie View A&M University. His research interests are in the Network Security, Cryptography and Aeronautical

Telecommunications Networks: Advances, Challenges, and Modeling, including artificial intelligence, multidiscipline advanced development technology management, and engineering software development.

429h Customization of an Entropy Generation Minimization Block in Aveva Process Simulation

Waleska Nascimento, Fernando Lima, João Manzi, Heleno Bispo

Entropy generation minimization (EGM) has been recognized as a useful approach to optimize process design and chemical industry operations. EGM has been applied in different fields, such as heat exchanger network synthesis, refrigeration cycle design, and power plant performance analysis (Linnhoff and Hindmarsh, 1983; Rao et al., 2003). The application of EGM has expanded to other areas, including chemical process design, control, and optimization (Manzi et al., 2011; Bispo et al., 2013; Maximo et al., 2019). Recently, a study by Bispo and Manzi (2022) showed the equivalence between minimum entropy and exergy applied to CSTRs, where optimal conditions have been determined and applied in a commercial simulator resulting in superior performance when compared to classical optimization procedures.

The present study describes the customization of a thermodynamic evaluation block in AVEVA Process Simulation (APS) based on the application of the concept of entropy generation minimization, creating the EGM block. APS is a widely-used software for process simulation, optimization, and process design. Specifically, the approach for constructing the EGM block involves the use of the Model Writing mode in APS for the reduction of the system's entropy generation rate in order to define optimal operating conditions, demonstrating the effectiveness of EGM in improving the efficiency and performance of chemical processes. The production of propylene glycol by the hydration of propylene oxide in a Continuous Stirred-Tank Reactor (CSTR) is used as the case study to demonstrate the proposed approach.

Several factors such as reaction temperature, pressure, residence time, and catalyst type have been investigated to optimize the process parameters and improve the product yield. So, in this work, the process of optimization has taken into account the mass, enthalpy, and entropy balances, allowing the entropy generation rate evaluation, as presented by Eq. (1). Assuming Eq. (1) is differentiable and a function of temperature (T), then, based on the more traditional optimization technique, the minimum value of it is obtained when the relationship, given by Eq. (2) is satisfied.

Based on Model Writing mode of APS, and taking into account the entropy generation rate minimization, a customized block, in Figure 1, has been created and used as an objective function. Then, the model block is used to calculate the entropy generation rate, considering the relation presented by Eq. (2), where the variables (Inlet temperature, residence time, and reaction temperature) are automatically adjusted to minimize the entropy generation rate. Such a procedure has allowed the process conversion improvement from 36% (Classical conditions - Case 01) to 95% (EGM conditions - Case 02), as presented in Figure 2.

Keywords: Entropy Generation Rate; Optimization; Thermodynamics.

References

Linnhoff, B., Hindmarsh, E. (1983), Chemical Engineering Science, 38 (5), 745-763

Rao, K. S., Reddy, K. S., Reddy, K. N. (2003), Energy Conversion and Management, 44(17), 2745-2758

Manzi, J., Brito, R., Bispo, H. (2011), Chemical Engineering Science, 66(3), 530-537

Bispo, H., Silva, N., Brito, R., Manzi, J. (2013), Energy Conversion and Management, 76, 26-31

Maximo, M., Bispo, H., Nilton, S., (2019), Chemical Engineering Communications, 206(9), 1241-1248

Bispo, H., Manzi, J. (2022), Chemical Engineering & Technology, 46(4), 785-793

$$\dot{\sigma} = - F^{e}\left(\sum C^{e}_{A_i} c_{p_{A_i}}\right)\left[\frac{(T - T^{e})}{T} + ln\left(\frac{T^{e}}{T}\right)\right] + rV\left(- \frac{\Delta G_R}{T}\right) \quad (1)$$

$$\frac{d\dot{\sigma}}{dT} = \left(T - T^{e}\right) + \frac{\tau k_0}{\Sigma\theta_i c_{p_i}}\{\frac{\left(\frac{E}{R}\right)e^{\left(-\frac{E}{RT}\right)}}{\left[1+\tau k_0 e^{\left(-\frac{E}{RT}\right)}\right]^2}\left(-\frac{\Delta\tilde{G}_r}{T_r} + \frac{\Delta\tilde{H}_r}{T_r}\right)$$

$$+ \Delta H_r\left[\frac{e^{\left(-\frac{E}{RT}\right)}}{\left[1+\tau k_0 e^{\left(-\frac{E}{RT}\right)}\right]} - \frac{\left(\frac{E}{R}\right)e^{\left(-\frac{E}{RT}\right)}}{T\left[1+\tau k_0 e^{\left(-\frac{E}{RT}\right)}\right]^2}\right]\} = 0 \quad (2)$$

Figure 1 - Customized APS Model Writing Block

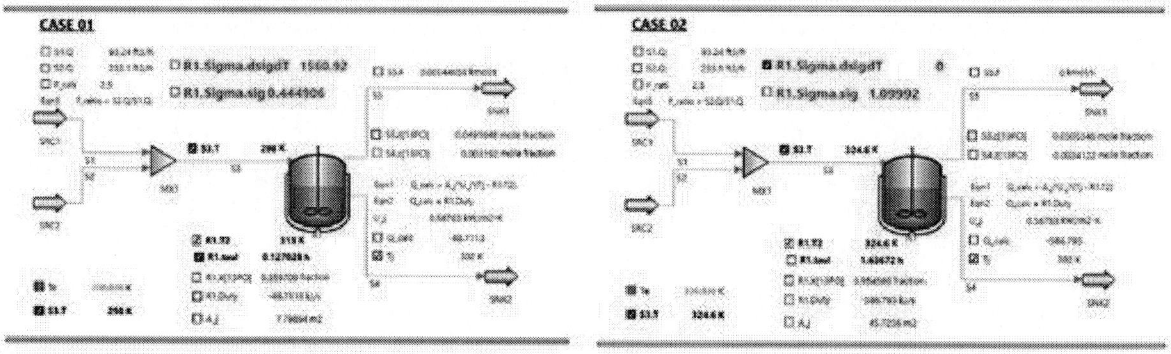

Figure 2 - Process Evaluation Results in APS

458a Glycopy: A Multiscale Model-Based Simulation, Optimization, and Optimal Control Python Package for Monoclonal Antibody Glycosylation in Cell Culture

Yingjie Ma, Jing Guo, Richard Braatz

Glycosylation – the controlled enzymatic modification of an organic molecule by addition of a sugar molecule – plays an important role in the clinical efficacy of monoclonal antibodies (mAb), the dominant products in the biopharmaceutical market [1]. As such, the glycosylation pattern is a critical quality attribute (CQA) that should be controlled at specific range to ensure product quality [2]. It is desirable to develop mathematical models to predict glycosylation profiles [3], which would then be used for analysis, design, and control as per the principles of quality-by-design (QbD) [4].

Several glycosylation models have been proposed in the literature [5–7]. The potential value of using such models in optimizing glycosylation has been demonstrated in a recent study [7], in which a model was used to increase the concentration of a galactosylated mAb by 93%. The objectives of our work are to further advance the optimization and control of glycosylation by (1) developing a multi-scale model for glycosylation that produces more accurate predictions, and (2) providing an open-source, flexible, and extendable package in Python [8] for the modeling, optimization, and optimal control of glycosylation, which we denote as *GlycoPy*.

Our multiscale model [7] consists of a cell culture model to predict mAb production rate, a nucleotide sugar donor (NSD) synthesis model to predict intracellular NSD concentrations, and golgi glycosylation (both N-linked and O-linked, steady-state and dynamic, simplified and original) models to predict the glycan profiles. The models are organized modularly so that the users can adapt and extend for their specific cells/cell lines conveniently and can include new mechanisms/submodels easily. Specifically, a reaction network module was defined to manage reaction models. Enzymes and transport proteins commonly used in different reactions are also submodels that can be defined independently. The package was built upon an automatic differentiation framework, so accurate derivatives can be generated efficiently, facilitating efficient and reliable model-based optimization. Since all the model variables and equations are symbols that are finally sent to integrators written in C language for dynamic simulation, the users do not need to worry about the inefficiency of Python language for solution. Based on the modeling platform, more functionalities including parameter estimation, dynamic optimization, model predictive control (MPC) (linear, nonlinear, and adaptive formulations), and optimal design of experiments (including batch and receding horizon methods) are developed with both single- and multiple-shooting algorithms implemented for solution.

The production of mAb from Chinese hamster ovary (CHO) cells was used as an example to demonstrate the functionalities of the package. The parameter estimation in this package uses automatic differentiation to increase the robustness and increase the speed by 1 to 2 orders of magnitude. Through applying the receding horizon method to design experiments online, we can estimate the model parameters (around 100 parameters) much more accurately than the best of the existing literature reports [e.g., Ref. 7]. Finally, adaptive MPC is shown to increase the mAb productivity with glycan profiles controlled within a desired range and outperform nonadaptive predictive controllers. The results demonstrate the efficiency, versatility, flexibility, and extendability of GlycoPy as an open-source package for modeling, optimization, optimal design of experiments, and optimal control of mAb glycosylation.

(1) Walsh, G. Biopharmaceutical Benchmarks 2018. *Nat. Biotechnol.* **2018**, *36* (12), 1136–1145.

https://doi.org/10.1038/nbt.4305.

(2) Batra, J.; Rathore, A. S. Glycosylation of Monoclonal Antibody Products: Current Status and Future Prospects. *Biotechnol. Prog.* **2016**, *32* (5), 1091–1102. https://doi.org/10.1002/btpr.2366.

(3) Majewska, N. I.; Tejada, M. L.; Betenbaugh, M. J.; Agarwal, N. N-Glycosylation of IgG and IgG-Like Recombinant Therapeutic Proteins: Why Is It Important and How Can We Control It? *Annu. Rev. Chem. Biomol. Eng.* **2020**, *11* (1), 311–338. https://doi.org/10.1146/annurev-chembioeng-102419-010001.

(4) Rathore, A. S.; Winkle, H. Quality by Design for Biopharmaceuticals. *Nat. Biotechnol.* **2009**, *27* (1), 26–34. https://doi.org/10.1038/nbt0109-26.

(5) Shelikoff, M.; Sinskey, A. J.; Stephanopoulos, G. A Modeling Framework for the Study of Protein Glycosylation. *Biotechnol. Bioeng.* **1996**, *50* (1), 73–90. https://doi.org/10.1002/(SICI)1097-0290(19960405)50:1<73::AID-BIT9>3.0.CO;2-Z.

(6) Ehsani, A.; Niedenfuehr, S.; Eissing, T.; Behnken, S.; Schuppert, A. How to Use Mechanistic Metabolic Modeling to Ensure High Quality Glycoprotein Production. In *Computer Aided Chemical Engineering*; Espuña, A., Graells, M., Puigjaner, L., Eds.; Elsevier, 2017; Vol. 40, pp 2839–2844. https://doi.org/10.1016/B978-0-444-63965-3.50475-X.

(7) Kotidis, P.; Jedrzejewski, P.; Sou, S. N.; Sellick, C.; Polizzi, K.; del Val, I. J.; Kontoravdi, C. Model-Based Optimization of Antibody Galactosylation in CHO Cell Culture. *Biotechnol. Bioeng.* **2019**, *116* (7), 1612–1626. https://doi.org/10.1002/bit.26960.

(8) Van Rossum, G.; Drake, F. L. *Python 3 Reference Manual*; CreateSpace: Scotts Valley, CA, 2009.

458b Towards 3-Fold Sustainability in Biopharmaceutical Supply Chains

Miriam Sarkis, Alasdair Fyfe, Kye Liew, Isabelle Rider, Andrea Bernardi, Jesslyn Fung, Ming Hei (Hale) Lee, Nilay Shah, Maria Papathanasiou

The biopharmaceutical industry is uniquely placed to create value for society, by improving disease outcomes. Recent innovations in targeted and personalized healthcare are paving the way to cure life-threatening diseases; as more therapies reach market approval, manufacturers should carefully orchestrate and expand resources to make supplies meet forecasted demands. In this context, capacity planning activities are driven towards manufacturing strategies that decrease therapy costs and maximize product availability to those in need.

Awareness of environmental sustainability of supply chain operations in the life science sector is becoming increasingly relevant [1]. Biopharmaceutical process steps are water and energy intensive while pollutants are emitted during distribution. Manufacturing platforms often rely on stainless steel or single-use equipment, which production and procurement activities also utilize resources and have associated environmental impacts [2]. During operation, selecting stainless steel would require the use of high-pressure steam for sterilization, whereby adopting single-use equipment translates in disposal of plastic waste at the end of each campaign or batch. Therapy manufacturers will face pressures from governmental agencies to improve sustainability and demonstrate green credentials of raw material suppliers, which highlights a need to quantify the environmental footprints of different manufacturing setups and minimize them where possible, thereby complying with the set 2030 Net Zero targets [3]. In this space, computer-aided decision-making can help assess trade-offs and design cost-effective supply chains (*economic* sustainability) with minimized environmental footprints (*environmental* sustainability), which ensure product availability for the population (*social* sustainability).

In this work, we demonstrate how 3-fold sustainability improvements in pharmaceutical manufacturing and distribution can be achieved, whereby economic, social, and environmental objectives are considered. We integrate the pharmaceutical supply chain investment and operational planning problem with life cycle assessment (LCA), focusing on emerging biopharmaceutical products (Figure 1). We present a mixed-integer linear programming (MILP) framework, which considers a set of candidate network nodes, demands, costs, operational capabilities, material intensity, transport emissions and indices and determines candidate supply chain structures, operational plans and selects suitable process equipment technologies.

Candidate supply chain structures are assessed with respect to the KPIs of (i) cost, (ii) product availability and (iii) environmental score, with the latter integrating environmental impacts related to emissions, water usage, human health, and energy consumption. The ε-constraint method is implemented to construct a Pareto frontier of candidate manufacturing setups. This enables insights into the trade-offs of financial and environmental sustainability and highlights opportunities brought by distributed manufacturing and selection of plastics-based single-use versus stainless steel equipment. A sensitivity analysis is performed to assess the impact of assumed weights and operational uncertainty on the results of the optimization.

References

[1] European Federation of Pharmaceutical Industries and Associations (EFPIA) (2022), Environment, Health, Safety and Sustainability

[2] World Steel Association (2020), Life cycle assessment in the steel industry

[3] HM Government (2021), Life Science Vision Report

Acknowledgements

Funding from the UK Engineering & Physical Sciences Research Council (EPSRC) for the Future Targeted Healthcare Manufacturing Hub hosted at University College London with UK university partners is gratefully acknowledged (Grant Reference: EP/P006485/1). Financial and in-kind support from the consortium of industrial users and sector organizations is also acknowledged.

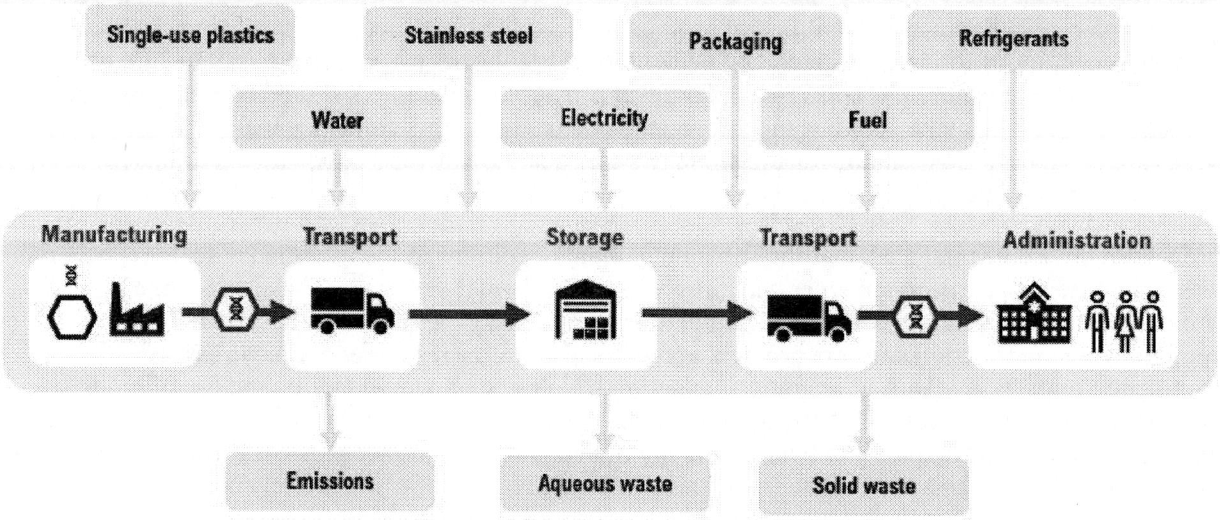

Figure 1: Key inputs and outputs of biopharmaceutical supply chains

458c Kinetic Modeling and Operability Analysis for the Optimization and Advanced Control of Xanthan Gum Production

Davi Oliveira, Ronald Alexander, Antonio C. L. Horta, Adilson Silva, Fernando Lima

Xanthan gum is a highly viscous biopolymer produced by bacteria from the genus *Xanthomonas* with several applications in industries such as for food and cosmetics production, and oil extraction [1]. However, the high viscosity culture broth poses major obstacles for the production of xanthan gum. In particular, challenges arise due to the poor control of process mass transfer and oxygen supply, which ultimately result in process efficiency issues [2-3]. The conventional strategy for controlling oxygen supply in stirred-tank reactors is based on measuring the dissolved oxygen concentration and manipulating the impeller rotation speed, air and oxygen flows. Normally, bioreactor monitoring systems are already equipped with PID (Proportional-Integral-Derivative) controllers for this purpose [4-5]. Nevertheless, there is an inherent limitation of this method related to the biological and rheological conditions, which produce a heterogeneous culture medium and unstable delays in sensor signals. For example, at certain product concentrations, the system starts to show instability in the controlled and manipulated variables. Motivated by these challenges, this work addresses the development of a kinetic modeling and operability analysis framework for the optimization and advanced control of xanthan gum production.

Modeling the behavior of the process (assuming non-Newtonian broth behavior), enables the model-based optimization and control of the production of the Xanthan gum biopolymer. In particular, a kinetic model is developed for Process Operability and Model Predictive Control (MPC) purposes describing the evolution of biomass, product, substrate, viscosity, and oxygen consumption, for oxygen transfer rate control. In this approach, the culture broth rheology can be controlled and maintained by manipulating the feed and broth outlet rates of a CSTR, air, and O_2 inlet to keep the oxygen and substrate fluxes within their optimal ranges. Specifically, process operability provides a better understanding of the feasible operating regions during the design stage to help in the controllability assessment of nonlinear systems, including bioprocesses [6-8]. In this method, a feasible working region for the controller can be defined by the achievable output set (AOS), considering the kinetic model and the process inputs within the available inputs set (AIS). This AOS region can then be compared with the desired output set (DOS) that includes desired indicators, such as dissolved oxygen and broth viscosity, and a desired input set (DIS) is defined to satisfy process goals [9-10]. As motivated above, an advanced control algorithm is required to stabilize the operating conditions due to the complex nonlinear nature of this process. Results of the process simulation applying the operability analysis will be discussed toward the future application of the MPC.

[1] Rosalam S, England R. Review of xanthan gum production from unmodified starches by Xanthomonas comprestris sp . 2006;39:197–207.

[2] Amanullah A, Tuttiett B, Nienow AW. Agitator speed and dissolved oxygen effects in xanthan fermentations. Biotechnol Bioeng. 1998;57(2):198–210.

[3] García-Ochoa F, Castro EG, Santos VE. Oxygen transfer and uptake rates during xanthan gum production. Enzyme Microb Technol. 2000;27(9):680–90.

[4] Alford JS. Bioprocess control: Advances and challenges. Comput Chem Eng. 2006;30(10–12):1464–75.

[5] Diaz C, Dieu P, Feuillerat C, Lelong P, Salome M. Adaptive predictive control of dissolved oxygen concentration in a laboratory-scale bioreactor. J Biotechnol. 1995;43(1):21–32.

[6] Mesquita TJB, Campani G, Giordano RC, Ribeiro MPA, Horta ACL, Zangirolami TC, Lima FV. Operability and biomimetic control of a micro-aerated fermentation process. Comput Chem Eng. 2021;155:107511. Available from: https://doi.org/10.1016/j.compchemeng.2021.107511

[7] Demirel SE, Li J, Hasan MF. Systematic process intensification. Curr Opin Chem Eng. 2019;25:108–13. Available from: https://doi.org/10.1016/j.coche.2018.12.001

[8] Svensson, E., Eriksson, K., Wik, T. Reasons to apply operability analysis in the design of integrated biorefineries. Biofuels, Bioprod. Biorefining. 2015; doi:10.1002/bbb.1530.

[9] Gazzaneo V, Carrasco JC, Vinson DR, Lima FV. Process Operability Algorithms: Past, Present, and Future Developments. Ind Eng Chem Res. 2020;59(6):2457–70.

[10] Gazzaneo V, Lima FV. Multilayer Operability Framework for Process Design, Intensification, and Modularization of Nonlinear Energy Systems. Ind Eng Chem Res. 2019;58(15):6069–79

458d Screening of Parameters Influencing Performance of Tangential/Crossflow Filtration Systems in Continuous API Manufacturing Processes with a Mechanistic CFD Model

Bennie Anderson, Fernando Muzzio, Ravendra Singh

The latest revision of Annex 1 has had many impacts on the pharmaceutical industry since its release, but one of the most significant changes is increased scrutiny upon sterilized filtration systems with the Pre-Use Post Sterilization Integrity Testing (PUPSIT) requirement [1]. PUPSIT requires pharmaceutical manufacturers to test sterilized filter assemblies prior to use by performing non-destructive integrity testing such as bubble point, diffusive flow, water intrusion or pressure hold testing, which (while not uncommon) was not previously standard practice across the industry [2]. As a result, this revision will impose significant difficulties upon some pharmaceutical manufacturers and require adjustments to their current manufacturing processes to remain in compliance with Good Manufacturing Practice (GMP) [2].

However, in cases where PUPSIT is not possible due to process constraints, alternative approaches may be acceptable, provided that a thorough risk assessment is performed, and compliance achieved through implementation of appropriate controls to mitigate risk in non-integral filtration systems. Per Annex 1, this risk assessment should include "in depth knowledge and control of the filter sterilization process to ensure the potential damage to the filter is minimized" [1].

The latest revision of Annex 1 was finalized in August 2022 and will be coming into effect in August 2023, one year after publication (excluding lyophilization). As the deadline grows closer, it is an especially critical time to investigate and better understand the multitude of factors influencing filtration systems in order to perform effective risk analysis and develop proper mitigation strategies.

One form of filtration showing considerable interest in recent years is cross flow filtration, also known as tangential flow filtration (TFF), which is in part due to its compatibility with continuous manufacturing processes and flow chemistry [3]. A major benefit of TFF systems relative to standard dead-end filtration is reduced buildup of particulate on the membrane's surface (the "filter cake") which is detrimental to effectiveness and can block a filter. In a TFF setup, the flow tangential to the membrane effectively "washes away" the particulate buildup and extends filter lifetime, which is ideal when employing continuous manufacturing of APIs.

In this work, a mechanistic CFD model of a TFF system is developed, calibrated and verified with experimental data. The model verification involved comparing the model prediction with the collected experimental data (residence time distribution, saturation rates, hold times, etc.) at various conditions which encompassed the standard operational range of said filtration unit. This model was then used to perform screening experiments with various material and operational parameters that impact the performance of a filter. Some system parameters used in screening include membrane thickness, permeability, fluid viscosity, and diffusion coefficients.

In a traditional lab setting, such an investigation would require many unique experimental setups, various materials, and likely utilize a Design of Experiments approach (DoE) with a limited number of runs as a fractional factorial to minimize time and cost. These fractional factorial approaches, while efficient, can limit the usefulness of generated statistical models. Without sufficient datapoints spanning the design space, interactions between parameters cannot be fully investigated and may lead to important interactions being missed [4].

However, with a mechanistic CFD model, the restrictions on the number of runs which accompany physical experiments are reduced and a full factorial experiment is possible. Upon completion of the screening runs, analysis was performed using statistical software to create various empirical linear and nonlinear models which are useful for understanding the complexities and interactions of the various parameters.

The primary objective of this study is to enhance our understanding of TFF/CFF filtration systems and demonstrate a method that can be widely applied across the pharmaceutical industry as the latest revision of Annex 1 comes into effect. This approach can be performed on other types of filtration systems and geometries within pharmaceutical manufacturing and other industries. The approach is simple and straightforward, which allows it to be easily adapted for screening experiments with more parameters under investigation.

References

[1] The Rules Governing Medicinal Products in the European Union Volume 4 EU Guidelines for Good Manufacturing Practice for Medicinal Products for Human and Veterinary Use, Annex 1: Manufacture of Sterile Medicinal Products. Brussels 22.8.22 C(2022) 5938 final https://health.ec.europa.eu/system/files/2022-08/20220825_gmp-an1_en_0.pdf

[2] 2022 Annex 1 Workshop (Palm Springs), Oct 20-21 2022, hosted by the Parenteral Drug Association.

[3] Huter, M; Strube, J. (2019). Model-based design and process optimization of continuous single pass tangential flow filtration focusing on continuous bioprocessing. Processes, 7(6), 317. https://doi.org/10.3390/pr7060317

[4] Hessing, T. (2023, March 25). Partial/fractional factorial design. Six Sigma Study Guide. Retrieved March 31, 2023, from https://sixsigmastudyguide.com/partialfractional-factorial-design/#:~:text=One%20of%20the%20big%20drawbacks,are%20confounded%20with%20other%20effects.

Acknowledgements

This work is supported by the US Food and Drug Administration (FDA) under contract number 75F40122C00122.

458e A Dynamic Hybrid Modeling Framework for Bioprocesses

Haiting Wang, Cleo Kontoravdi, Antonio del Rio Chanona

One of the main hurdles in the computer-aided design and optimization of industrial bioprocesses is the limited capability of models to accurately represent biological systems. While mechanistic models are hindered by the definition and implementation of all the potentially relevant biomechanisms, data-driven models are limited by their inability to extrapolate. Hybrid models can present the best of both worlds. A well-built hybrid model can be prototyped relatively quickly, and provide reliable predictions with good extrapolation capabilities[1]. Unfortunately, building a 'correct' hybrid model, which accurately balances its data-driven and first principles components, is not trivial.

This work introduces a novel framework for constructing hybrid models, which integrates a mechanistic backbone with data-driven components. The mechanistic backbone is informed by prior knowledge of biological mechanisms, while the data-driven components are incorporated through statistical model selection methods to find the 'correct' hybrid model that balances the fitting accuracy and extrapolation ability. Statistical model selection methods such as the Bayesian Information Criterion (BIC)[3], the Akaike Information Criterion (AIC)[4], and the Hannan Quinn Criterion (HQC)[5] are embedded into the framework. When the statistical methods identify different "correct" model structures, a further optimal experimental design is completed to discriminate between the selected structures.

Two bioprocess case studies are considered in this work with different system complexity. Firstly, the proposed framework is applied in a simple biomass cultivation case study. Two hybrid models have been constructed, one by applying polynomial regression (hybrid–polynomial) and the artificial neural networks (hybrid–ANNs) respectively. The results demonstrate the robustness of our framework against high noise.

The framework is further tested on a microalgae cultivation process for biodiesel production[6]. Following the same procedure, the mechanistic structure of the hybrid model is designed based on mass balances of state variables. The influence of light attenuation, substrate and nitrate consumption on the microalgae growth rate is modeled through polynomial regression and ANNs. Finally, the extrapolation capability of the statistically selected hybrid model is tested by varying the initial conditions of state variables and incident light intensity by $\pm$ 20% to stress-test the framework. It is found that both the hybrid–polynomial and the hybrid–ANNs can simulate the microalgae system with good extrapolation ability and avoid complex mechanistical expression of biomechanisms at the same time. One observation of this work is that the selected information criteria sometimes select different optimal model structures. This is not surprising due to their different statistical derivations. To account for this deviation, a further experimental design optimization is completed to further discriminate between the selected model structures.

In summary, we demonstrate how our proposed hybrid model construction framework can be applied for dynamic modelling for bioprocess. The framework is flexible regarding the system complexity and robust against high measurement noise.

References

[1] M. Von Stosch, R. Oliveira, J. Peres, and S. F. de Azevedo, 'Hybrid semi-parametric modeling in process systems engineering: Past, present and future,' *Computers & Chemical Engineering,* vol. 60, pp. 86-101, 2014.

[2] P. Kotidis and C. Kontoravdi, 'Harnessing the potential of artificial neural networks for predicting protein glycosylation,' *Metab Eng Commun,* vol. 10, p. e00131, Jun 2020, doi: 10.1016/j.mec.2020.e00131.

[3] A. A. Neath and J. E. Cavanaugh, 'The Bayesian information criterion: background, derivation, and applications,' *Wiley Interdisciplinary Reviews: Computational Statistics,* vol. 4, no. 2, pp. 199-203, 2012.

[4] Y. Sakamoto, M. Ishiguro, and G. Kitagawa, 'Akaike information criterion statistics,' *Dordrecht, The Netherlands: D. Reidel,* vol. 81, no. 10.5555, p. 26853, 1986.

[5] E. J. Hannan and B. G. Quinn, 'The determination of the order of an autoregression,' *Journal of the Royal Statistical Society: Series B (Methodological),* vol. 41, no. 2, pp. 190-195, 1979.

[6] E. A. del Rio-Chanona *et al.*, 'Dynamic modeling of green algae cultivation in a photobioreactor for sustainable biodiesel production,' *Biotechnology and bioengineering,* vol. 115, no. 2, pp. 359-370, 2018.

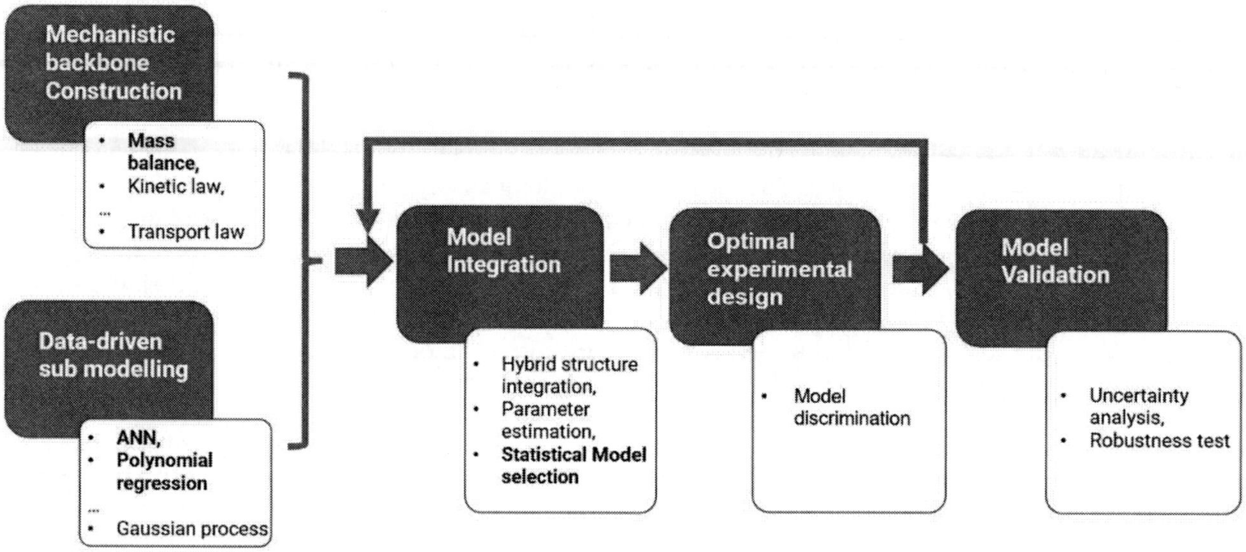

Figure 1 Illustration of the hybrid model construction framework.

458f A Dynamic Metabolic Flux Analysis (DMFA) Model for the Production of Monoclonal Antibodies in CHO Cell Perfusion Culture.

Nikola Malinov, Jayanth Venkatarama Reddy, Eleftherios Papoutsakis, Marianthi Ierapetritou

Monoclonal antibodies (mAbs) are a class of recombinant proteins that have revolutionized medicine since the 1980's providing life-saving therapies for a multitude of diseases.[1] Chinese hamster ovary (CHO) cells are the preferred mammalian cell platform for mAb production given their robust growth in defined mediums, easy transfectability, and native machinery which induce complex post-translational modifications.[2,3] Perfusion bioreactors facilitate continuous upstream cell culture and therefore process intensification by enabling high viable cell densities (VCDs), increased productivities, reduced equipment sizes, and improved product quality. However, the increased operational intensity relative to fed batch mode poses a significant challenge to attain optimal operation and control of perfusion bioreactors.

Mechanistic models are a valuable tool for the biopharmaceutical industry as they leverage the physical laws underpinning the process to link the key parameters with the Critical Quality Attributes (CQAs) of the product. Advancement of mechanistic models in biomanufacturing has also been heavily emphasized by the Food and Drug Administration (FDA) as a platform to implement the Quality by Design (QbD) process development principle.[4] With respect to upstream biopharmaceutical production, modeling cell metabolism is imperative to analyze nutrient demands and physiologic objectives under different operating conditions. Specifically for continuous perfusion culture, modeling CHO cell metabolism is in its infancy compared to the extensive efforts in the literature for fed batch.

In this work, a comprehensive Dynamic Metabolic Flux Analysis (DMFA) model is developed for CHO cell perfusion culture based on prior analysis of fed batch operation. The model accounts for the effects of the perfusion rate and the cell bleed rate on the bulk mass transport and cell growth dynamics of the system. The bleed rate is simulated as a discrete event to align with its manual intermittent use in the experimental data collected for model regression. The required inputs are the initial extracellular metabolite concentrations, media amino acid and glucose profile, process conditions, and a robust kinetic parameter set. The outputs include simulated extracellular concentration profiles, intracellular and exchange fluxes, and the production and growth rates. The framework consists of a network of 67 biochemical reactions and 42 metabolites representing the key metabolic pathways in CHO cells while maintaining an overdetermined system and thus allowing application of MFA to estimate the intracellular fluxes. The pathways include glycolysis, the TCA cycle, amino acid metabolism, oxidative phosphorylation, the urea cycle, and biomass and antibody synthesis. Metabolite mass balances under the pseudo-steady state assumption yield the stoichiometric component of the model. The kinetic component has been developed to include Monod expressions describing the exchange fluxes as a function of the extracellular metabolite concentrations.

Experimental data from the CHO VRC01 cell line grown in perfusion culture has validated the model's application to perfusion and allowed estimation of the kinetic parameters. The experimental platform consisted of a 1 L vessel controlled by the Eppendorf BioFlo 120 system. The cell retention device was a Spectrum® Hollow Fiber Filter Module from Repligen with a 0.65 μm pore size operated in tangential flow filtration mode. A manual, intermittent cell bleed was used to control the VCD to within a $\pm 4\times10^6$ cells mL^{-1} deadband about the setpoint of 20×10^6 cells mL^{-1}. The perfusion rate was set to 1 vvd^{-1} while all other process parameters were maintained at industrially relevant values typical for CHO cells. The media formulation was based on an in-house fed-batch recipe which blends commercially available basal and feed medias, along with supplemental glucose. Future work seeks to improve the mathematical framework developed here with additional experimental data and subsequently apply it towards perfusion process optimization objectives, such as metabolic control and media reformulation.

References:

[1] Ecker, D. M., Jones, S. D., & Levine, H. L. (2015, January). The therapeutic monoclonal antibody market. In *MAbs* (Vol. 7, No. 1, pp. 9-14). Taylor & Francis.

[2] Trill, J. J., Shatzman, A. R., & Subinay, G. (1995). Production of monoclonal antibodies in COS and CHO cells. Current Opinion in Biotechnology, 6(5), 553-560.

[3] Wurm, F. M. (2004). Production of recombinant protein therapeutics in cultivated mammalian cells. Nature biotechnology, 22(11), 1393-1398.

[4] Yu, L. X., Amidon, G., Khan, M. A., Hoag, S. W., Polli, J., Raju, G. K., & Woodcock, J. (2014). Understanding pharmaceutical quality by design. The AAPS journal, 16(4), 771-783.

458g Automatic in-Line Electrode Fault Detection Via Optical Coherence Tomography

Elisabeth Fink, Johannes Poms, Daniel Unger, Matthias Wolfgang, David Cantillo

Electrochemical processes are an important part of the production of active pharmaceutical ingredients (APIs). Faults of the electrode, such as corrosion or fouling, have a significant impact on the process performance. In severe cases, they can damage the electrode beyond repair and halt the conversion. In-line or operando monitoring of electrochemical processes have been established as an important link to process performance [1-6]. In the present work, a new method for direct electrode monitoring via optical coherence tomography (OCT) is presented. Due to the low wavelength interferometry that OCT is based on, some materials, such as particles, can be penetrated and will appear semi-transparent in the resulting 3D volume scans. On the other hand, corrosion bubbles as little as 20 μm in height on the electrode are already visible on the images. During the electrochemical process, scans of the surface of the electrode are taken via OCT and the resulting images are automatically evaluated. During the evaluation, the upper boundary of the surface is extracted and an idealized mathematical extraction model is computed. This enables the automatic classification of irregularities on the electrode, such as corrosion peaks or particles sticking to the electrode. The developed OCT technology allows for continuous monitoring and automatic real-time evaluation of the resulting images. Hence, the results from OCT can be used for a real-time control strategy aimed at improving and optimizing the electrochemical process.

[1] Star AG, Fuller TF. FIB-SEM Tomography Connects Microstructure to Corrosion-Induced Performance Loss in PEMFC Cathodes. *J Electrochem Soc*. 2017;164(9). doi:10.1149/2.0321709jes

[2] Menzel, J., Slesinski, A., Galek, P., Bujewska, P., Kachmar, A., Frąckowiak, E., Washio, A., Yamamoto, H., Ishikawa, M., & Fic, K. (2022). Operando monitoring of activated carbon electrodes operating with aqueous electrolytes. *Energy Storage Materials*, *49*, 518–528. https://doi.org/10.1016/J.ENSM.2022.04.030

[3] Wang, R., Zhang, H., Liu, Q., Liu, F., Han, X., Liu, X., Li, K., Xiao, G., Albert, J., Lu, X., & Guo, T. (2022). Operando monitoring of ion activities in aqueous batteries with plasmonic fiber-optic sensors. *Nature Communications 2022 13:1*, *13*(1), 1–11. https://doi.org/10.1038/s41467-022-28267-y

[4] Shan, X., Patel, U., Wang, S., Iglesias, R., & Tao, N. (2010). Imaging local electrochemical current via surface plasmon resonance. *Science*, *327*(5971), 1363–1366. https://doi.org/10.1126/SCIENCE.1186476

[5] He, M., Fic, K., Frackowiak, E., Novák, P., & Berg, E. J. (2016). Ageing phenomena in high-voltage aqueous supercapacitors investigated by in situ gas analysis. *Energy & Environmental Science*, *9*(2), 623–633. https://doi.org/10.1039/C5EE02875B

[6] Michel, L., & Ueli Angst, M. (2022). Automated local electrochemical characterization on metals with complex shape and practice-related surface state. *Measurement*, *201*, 111713. https://doi.org/10.1016/J.MEASUREMENT.2022.111713

458h Reducing Waste and Energy Consumption in Food Processing Industries: Two Applications

Mrunal Sontakke, Sambit Ghosh, Craig Dory, Alan Ganz, Henry Weber, Joel Plawsky, Johnson Samuel, B. Wayne Bequette

The food processing industries are energy intensive and, without proper monitoring, might lead to enormous energy wastage. The current study addresses this challenge through two viewpoints: reducing material waste through online quality management and reducing energy consumption through modeling techniques. These methods can be applied to Small and Medium-scale food industries by simplifying Smart manufacturing and helping them utilize the benefits of Industry 4.0.

Case 1: The first case views the problem from the perspective of reducing material wastage through online quality management using cost-effective, easy-to-deploy sensors that can provide quantitative detection of the "feel" of the product. The developed sensors involve an innovative application of the fiber-optic-based optical analyzer and Resistive Power Draw sensors. Enterprise-friendly composite data parameters are being developed for intuitive quality measures compared to the complex spectra from the optical analyzer.

Case 2: The second case studies the problem from the perspective of reducing energy consumption during the cooling and ripening of nutrition bars using coolant. The project aims at developing predictive cooling models based on the composition of the material, with an objective function of reducing energy consumption per unit of product.

With the help of the Smart Manufacturing and Innovation Platform (SMIP), developed by the Clean Energy Smart Manufacturing Innovation Institute (CESMII), data collection, sharing, and storage have become more manageable. This eases the task of implementing the new strategies.

425a Fabrication Form and Function

Pierre-Thomas Brun

This talk accounts for recent progress in the directed control of interfacial fluid flows harnessed to assemble architected soft materials. We are focusing on the paradigmatic problem of free-surface flows in curable elastomers. These elastomers are initially liquid and cure into elastic solids whose shape is imparted by concomitant and competing phenomena: flow-induced deformations and curing. Particular attention is given to the role of capillary forces in these systems. Originating from the cohesive nature of liquids and thus favoring smooth interfaces, capillary forces can also promote the destabilization of interfaces, e.g., into droplets. In turn, such mechanical instabilities tend to grow into regular patterns, e.g., hexagonal lattices. We discuss how the universality, robustness, and ultimate regularity of these out-of-equilibrium processes could serve as a basis for new fabrication paradigms, where instabilities are directed to generate target architected solids obtained without each element laid in place by mechanized intervention.

425b Compatible Monomers for Ring-Opening Metathesis Polymerization Additive Manufacturing

Alex Commisso, Samuel Leguizamon

Stereolithographic additive manufacturing (SLA) processes provide agility and complex design development of components necessary for a wide range of critical applications. Yet this technique is limited to a small subset of monomers that traditionally produce polymer materials with poor thermal and mechanical properties. Conversely, ring-opening metathesis polymerization (ROMP) can produce relatively tough, high T_g thermosets using latent ruthenium catalysts. Furthermore, spatiotemporal control of the polymerization using photosensitizers affords high resolution patterning of high T_g thermosets using photoROMP. Dicyclopentadiene (DCPD), an inexpensive but robust photoROMP monomer, has been featured in SLA resins to generate highly complex three-dimensional geometries at print speeds approaching industry-leading standards using continuous SLA techniques. Excluding DCPD, the monomer space for photoROMP and continuous SLA resins has been relatively underexplored. In this work, we synthesized and evaluated the potential of a library of strained olefins for generating robust polymer materials using photoROMP and continuous stereolithography.

425d 3D Printing of Reactive Porous Media: Impact of Surface Functionalization on Mineral Growth Kinetics

Harrish Kumar Senthil Kumar, Abdullah Al Nahian, Lauren Beckingham, Bryan Beckingham

There is a growing need to reduce the level of carbon dioxide (CO_2) in the atmosphere as the increasing atmospheric CO_2 has led to increases in global temperatures and other climate change impacts. Geological carbon sequestration aims to store captured CO_2 in deep geological formations. Therefore, understanding the rates and mechanisms of geochemical reactions is vital in these subsurface systems when exposed to CO2 is critical. As natural samples are heterogeneous and non-duplicative, we investigate methods for fabricating synthetic replicates of reactive rock samples using 3D printing to assess the impact of mineral dissolution and precipitation on porosity and permeability. X-ray Micro CT images of natural samples was used to generate 3D models for fused filament fabrication 3D printing of high impact polystyrene. Surface functionalization was applied to sulfonate the 3D printed surfaces, where the sulphonic acid moieties will subsequently seed calcite growth. FTIR spectroscopy was employed to confirm the surface presence of sulfonic acid groups. Precipitation of calcite crystals on 2D reactive HIPS films was evaluated gravimetrically, by X-ray Micro-CT and by XRD analysis. Weight based precipitation experiments show gradual increase in the calcite deposition on the 2D films over time. To analyze the effect of surface functionalization, the 2D films were surface functionalized over time and calcite deposition flow experiments were performed and confirmed that increase in surface functionalization facilitates improved deposition of calcite on 2D films. This approach is extended to 3D printed core samples and batch experiment showed internal precipitation of calcite as confirmed by XRD. Overall, this study helps to evaluate the reactive rock sample replicate's mineral reaction and precipitation within the porous structures using 3D printing, mimicking geochemical reactions from geochemical systems, and facilitating the understanding of geological reactions caused by CO_2 storage in sub-surface systems.

425e Analysis of Therapeutic Drug Release from Additively Manufacturable Materials As Affected By Mechanical Stress and Strain

Leyton Janowsky, Edward W. Davis

The release profile of a model therapeutic agent from a drug eluting biodegradable and additively manufacturable polymer was assessed under cyclic mechanical loading. Therapeutic agents within implanted medical components can serve a wide variety of useful purposes from enhancing the performance of artificial joints to preventing thrombosis around a stint. However, little data exists at this time on the effect a repetitive mechanical loading may have on the release profile of the therapeutic agent. While a stint may be regularly loaded at frequencies similar to 1 Hz, other medical components, such as artificial joints, may experience completely different mechanical environments with irregular, largely spaced loadings. Such wide differences in environment highlight the need to understand the possible effects mechanical loadings can have on the release from these components. This work is intended to provide a more complete understanding of these materials by collecting release profile data, under a variety of mechanical conditions, for an SLA printable poly(ethylene glycol) diacrylate-based biodegradable material containing aspirin as a model therapeutic. Cyclic loads, of frequencies from 0.1 to 2.0 Hz and strains from 1 to 5 % over the course of 240 hours, were applied while the sample was submersed in a phosphate buffer. Samples of dissolution media were extracted, and their aspirin content determined, 4 times every 24 hours to track release rates while DMA frequency sweeps were performed in the same intervals to track the material's degradation rate. The results serve to quantify the role of mechanical loading upon the release profiles from these materials, as well as the interaction between these properties and the decomposition of the sample.

425f Centrifugal Force Spinnning and Spinnability of Polymer Solutions

Cheryl Slykas, Carina Martinez, Louie Edano, Jorgo Merchiers, Naveen Reddy, Vivek Sharma

Designing application-ready fibers involves multifaceted challenges related to correlating the formulation properties and processing parameters to the fiber engineering trifecta of spinnability, morphology, and properties. Here, we characterize the influence of macromolecular and solvent properties on the trifecta for poly(ethylene oxide) or PEO fibers produced using a bespoke centrifugal force spinning (CFS) set-up and matched processing parameters. We illustrate the influence of changing solvent on spinnability, morphology, and properties (thermal and mechanical) by varying acetonitrile (AcN) fraction in the spinning dope formulated with PEO dissolved in AcN:H_2O mixtures. We contrast the numerical values of measured diameter, tensile strength, elongation-at-break, and crystallinity of centrifugally spun PEO fibers with the published datasets for electrospun fibers using the Berry number (or the overlap parameter) as the ordinate. We compile, analyze, and replot ES and CFS spinning datasets obtained for various solvents, PEO (M_w and c), and processing parameters. Even though distinct forces determine the jet trajectory and fiber formation for ES and CFS, we find centrifugally spun PEO fibers emulate electrospun fiber properties, morphology, and spinnability. We discuss the mechanism underlying volatile-entangled (VE) spinnability, displayed here by PEO solutions in certain AcN:H_2O mixtures, in contrast to extensibility-enriched (EE) spinnability of multicomponent formulations, enabled by the addition of an ultrahigh M_w polymer fraction.

AUTHOR INDEX

AIChE
120 Wall Street, FL 23
New York, NY 10005-4020

ISBN 978-1-7138-9293-9

CURRAN ASSOCIATES INC.
proceedings.com